ABC OF CAPACITORS

ABC OF CAPACITORS

BASIC PRINCIPLES, CHARACTERISTICS AND CAPACITOR TYPES

IMPRINT

PUBLISHER
Würth Elektronik eiSos GmbH & Co. KG
Max-Eyth-Str. 1 · D-74638 Waldenburg
Tel. +49 7942 945-0 · Fax +49 7942 945-5000
eiSos@we-online.com · www.we-online.com

IMPLEMENTATION
Anna Rudolf

AUTHOR
Stephan Menzel

GRAPHIC DESIGN // DIE NECKARPRINZEN · D-74074 Heilbronn
TYPESETTING // Ideen im Kopf · D-75050 Gemmingen
PRINT // E&B engelhardt und bauer · D-76131 Karlsruhe
PUBLISHING HOUSE // Swiridoff Verlag · D-74653 Künzelsau

EDITION
1st edition November 2014

ISBN 978-3-89929-294-7

Contents

Preface

Dear readers,

It is important for Würth Elektronik eiSos GmbH & Co. KG to provide technical knowledge and to share it with customers. Technical knowledge has already been published in the form of books covering the fields of inductors, transformers, electromechanical components and simulation with LTspice.

While successively expanding our portfolio with the field of capacitors, we would now also like to pursue this successful path together with our customers.

It was our wish to provide the reader with a book that summarizes the essential basics of capacitors and to convey this knowledge in an easy and accessible way.

The idea has now been realized. The Abc of Capacitors is a reference book covering the relationships and technical characteristics of capacitors in a condensed form.

For us, the Abc of Capacitors represents the first step towards communicating technical knowledge, application and deployment options covering all aspects of capacitors. Further publications on the application and deployment options for capacitors are lined up for the future.

We hope that you enjoy reading this book.

Alexander Gerfer

Director International Product Management

Würth Elektronik eiSos GmbH & Co. KG

Stephan Menzel

Product Development Manager Capacitors

Würth Elektronik eiSos GmbH & Co. KG

The author

Stephan Menzel, born in 1987 and after graduating in industrial engineering from the Reinhold-Würth-Hochschule, worked at Würth Elektronik eiSos in Quality Management. In his role as a Quality Manager, he was responsible for standardization and supported key account customers. Following his master degree from the University of Wismar for which he studied alongside employment, he switched to the Product Development Management department where he has since lead the capacitor business development. In his job role he is responsible for aluminum and polymer-electrolytic capacitors.

Copyright

Suggestions and criticisms on this book gratefully received by:

Würth Elektronik eiSos GmbH & Co. KG · Mr. Stephan Menzel · e-mail: stephan.menzel@we-online.de

Tel. +49 7942 945-0 · Fax +49 7942 945-5009

Abc of Capacitors

Basic Principles

Basic principles, characteristics and capacitor types

1 Basic Principles

1.1 Electric field

In order to understand the conditions in the electric field, some relationships need to be explained. These serve to provide a basic understanding and are necessary to be able to translate the relationships to the capacitor application. The journey into the world of capacitors begins with the structure of the atom.

The structure of an atom consists of protons, electrons and neutrons. The electron is the carrier of the smallest possible negative electric charge (–e). The charge of an electron corresponds to the magnitude of the elementary charge (e):

$$e = 1.602 * 10^{-19}\ \mathrm{As} \tag{1.1}$$

The proton has the opposite charge to that of the electron, but of the same magnitude. The proton is therefore a carrier of positive charge (+e) of the same magnitude as the elementary charge. In contrast to protons and electrons, neutrons are electrically neutral and have neither a negative nor a positive charge.

The charge or charge quantity of a space is the difference between the quantity of positive charge and the quantity of negative charge present. The result is two types of electrical charge – positive and negative. Positive charge exists if the quantity of positive charge predominates. This state is known as a deficit of electrons. On the other hand, there is a negative charge if there is more negative than positive charge and the resulting difference is therefore negative.

This situation is termed as an electron surplus. It should be pointed out that every electric charge is always an integral multiple of the elementary charge e.

Every electric charge transforms the surrounding space into a special state. This special form of spatial state is termed electric field. Here the space, which is empty or filled with material, is under the influence of the separated electrical charges. The electric field is characterized by the charges exerting forces on other charges and therefore a certain force acts on electrically charged bodies.

If the electric field is caused by stationary charge, this is also termed an electrostatic field. Conversely, an electric field arising from moving charge is known as an electrodynamic field.

Electric field

1.2 Electric field strength

Electric field strength

The force acting on a charge in the electric field is known as the electric field strength. The electric field strength is a vector and is usually measured in N/As or V/m.

So-called field lines are used to illustrate and present an electric field. The field lines depict the direction of the electric forces acting for each point in the space. The field lines are always directed from plus to minus, so they always run from the positive to the negative charge.

The following figure shows an example of the force exerted on a charge in the field:

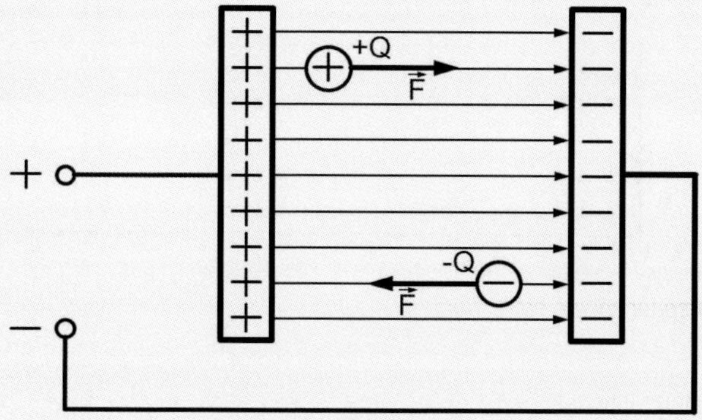

Fig. 1.1: Force exerted on charges in the electric field

The magnitude of the electrical field strength E is determined as follows and can be calculated from the force *F* applied and the charge *Q*:

$$E = \frac{F}{Q} \qquad [E] = \frac{N}{As} \tag{1.2}$$

If an electric field is of the same field strength at all points, it is known as a homogeneous electric field. In the homogeneous electric field the following relationship applies for the electric field strength E:

$$E = \frac{U}{d} \qquad [E] = \frac{V}{m} \tag{1.3}$$

In the case of a homogeneous electric field, all field lines run in parallel and have the same distance to one another (equidistant). The state of a homogeneous electric field is illustrated as follows:

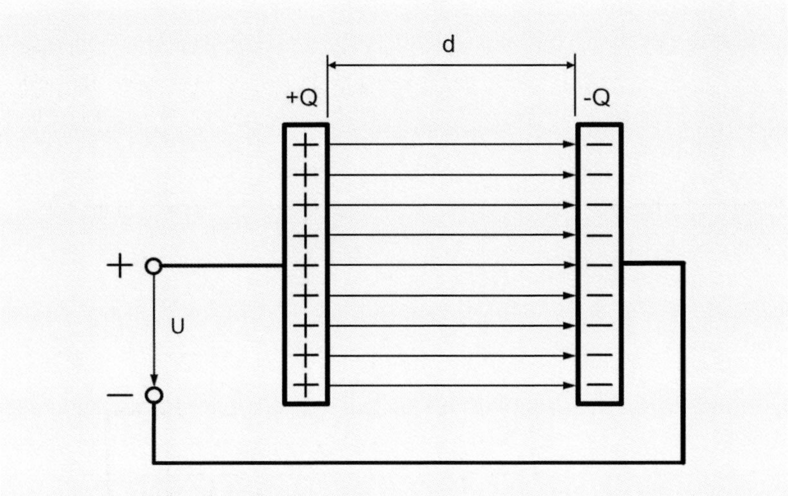

Fig. 1.2: Homogeneous electric field

In the case of a homogeneous electric field, the level of the electric field, as presented in Equation 1.3, is determined by the following factors:

- U ... Magnitude of the charge difference
- d ... Separation of the charged particles

So with a rise in the difference in charge, the size of the resulting electric voltage increases proportionally and therefore also the electric field strength. Furthermore, the electric field strength rises with a reduction in the separation between the charged plates.

If the field strength and/or the field direction depends on location and is not the same at every point in the field, as with a homogeneous field, the field is inhomogeneous.

1.3 Electrostatic induction

Electrostatic induction

The term electrostatic induction describes the temporary situation in which there is a shift in charge under the influence of an external electric field.

If a metallic conductor/metal body is introduced into an existing electric field, a shift in charge of the negative charge carriers results towards the positively charged electrode. This causes a shift of the freely moving electrons against the field direction and this is accompanied by a shift/separation of charge on the metallic conductor. The charge of the conductor introduced into the field then shifts to the side from which the electrons shift, to the positive.

If the existing external electric field is removed, the phenomenon is neutralized and the shift in charge disappears again.

The following represents states of an electric field with a conductor introduced. This illustrates the sequence of electrostatic induction:

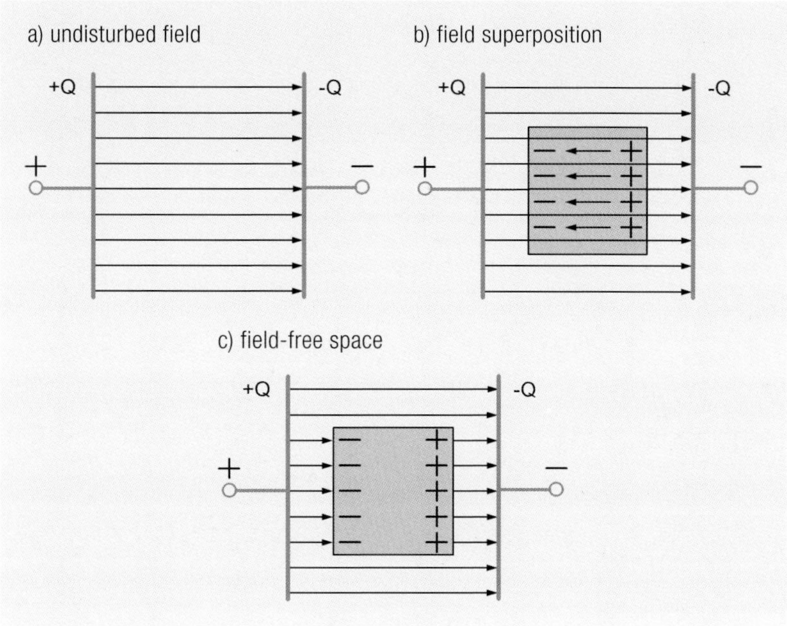

Fig. 1.3: Electrostatic induction through an electric field in a conductor

In state a) (see Figure 1.3) the undisturbed field without the conductor introduced is represented in the starting state. If now, as in b), a metallic conductor is introduced into the electric field, electrical forces act on the charges on the conductor. There is a shift of charge, as described, whereby there is a separation of charge inside the conductor and this forms another intrinsic electric field. As soon as the counter field of the conductor is the same as the external field, the shift in charge is concluded. In this state c) there is no field inside the conductor. The field-free space in the conductor is shielded against the external field and the field-free space is known as a "Faraday cage".

1.4 Electric dipoles

Considered externally, atoms and molecules have no electrical effect under normal conditions and are electrically neutral towards the outside. Through the influence of electrical fields or on the basis of internal binding forces, in the case of asymmetrical molecules, it may occur that the centers of charge of the positively charged atomic nuclei and the negatively charged shell become locally displaced or no longer coincide. Should this happen, the molecule then has a dipole moment. Consequently, the molecules are charged differently at their ends and molecules that show this effect are known as dipoles (or also dipole molecules).

Water is an example for the formation of a dipole. Here the dipole formation is highly pronounced, as the oxygen atom binds the two electrons from the hydrogen atoms.

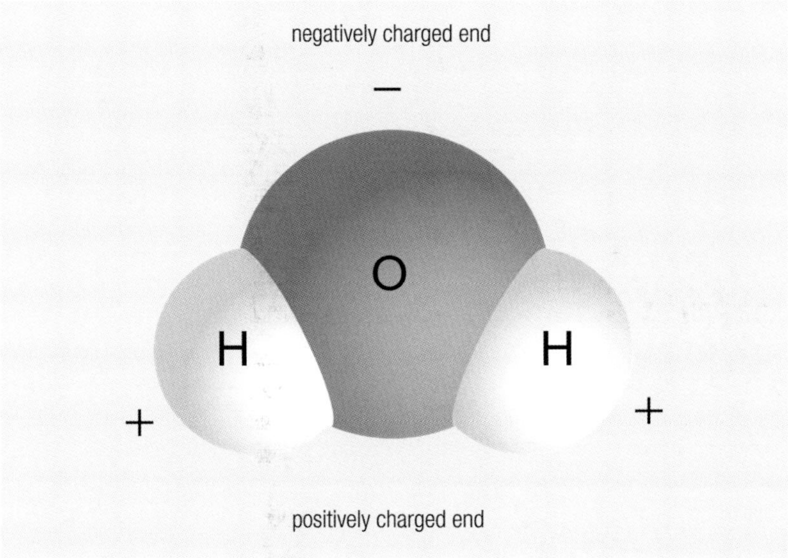

Fig. 1.4: Water molecule as an example of a dipole

1.5 Polarization

The shift in charge described in Chapter 1.3 Electrostatic induction not only exists in electrical conductors, but also in insulating materials. However, there is an essential difference in this case.

Shift in charge

In the electrical conductor there is movement of free (non-localized) electrons with the shift in charge. With insulating materials, however, the charge carriers are localized and externally the individual molecules are electrically neutral. The charges in an insulating material can only be shifted or rotated within their associated atoms or molecules by the influence of an external field. The atoms of the insulating material, which are actually neutral, become small electrical dipoles through the influence of the electric field.

The process in which an elastic shift in charge or rotation of the charges occurs in the insulating material is known as elastic or reversible polarization. The polarization contributes to the displacement flux ψ accordingly (see Chapter 1.7. Displacement flux and displacement flux density).

1.6 Dielectric and permittivity

Dielectric

A dielectric is a non-metallic insulating material, which has electrically weak to non-conductive properties. The charge carriers of a dielectric are not free to move and are localized on the molecular level. A dielectric describes the insulating material placed in the space between and exceptionally on two charged metal plates or in the electric field space.

Dielectrics are insulating materials used in electronic components. In the case of a capacitor, the dielectric is the actual effective insulating layer between the electrode surfaces of the capacitor (see Chapter 1.8 Capacitor).

Permittivity ε

A dielectric is characterized by the permittivity ε (formerly also known as dielectric constant). If, instead of air, another medium is used to insulate between the electrodes of the capacitor, the capacitance of the capacitor changes. The reason for this is the molecular structure of the dielectric. The associated localized charges of the molecules are small electric dipoles. Under the influence of the electric field, the dipoles orientate themselves according to the field (polarization) and the displacement flux ψ is amplified.

Permittivity denotes how the introduction of the dielectric changes the charge bound in the field. Absolute permittivity ε_0 and relative permittivity ε_r are distinguished. Permittivity ε is calculated by multiplying absolute and relative permittivity:

$$\varepsilon = \varepsilon_0 * \varepsilon_r \qquad (1.4)$$

Absolute permittivity ε_0

Absolute permittivity ε_0 is a natural constant and is also known as the electrostatic induction constant or electric field constant (permittivity of vacuum).

$$\varepsilon_0 = 8.8542 * 10^{-12} \frac{As}{Vm} = 8.8542 * 10^{-14} \frac{F}{cm} \qquad (1.5)$$

Relative permittivity ε_r

Relative permittivity ε_r is a material-specific quality criterion for the respective dielectric. The dielectric constant ε_r is specific to each dielectric. The specification does not have a unit, as it is a dimensionless number.

Relative permittivity ε_r denotes how many times better the dielectric is than vacuum (ε_r of vacuum = 1). The relative permittivity ε_r of air is 1.00058 (at a temperature of 0 °C and an atmospheric pressure of 101.3 kPa) and is usually approximated to 1 in practice.

The higher the relative permittivity ε_r, the higher the bound charge in the electric field. A material with a higher permittivity exhibits correspondingly better dipole formation.

Depending on the material, the relative permittivity is also highly dependent on the following factors:

- Frequency
- Temperature
- Voltage
- Time (there are dielectrics with aging behavior, e.g. class 2 ceramic capacitors)

The following table is an overview of the relative permittivity of different insulating materials, which are used e.g. in capacitors.

Material	Relative permittivity (ε_r) at 20 °C
Vacuum	1
Air (1 bar)	1.00059
Paper	1.6 … 2
Waxed paper	2
Polytetrafluorethylene (Teflon)	2.1
Polystyrene	2.3
Polypropylene	2.5
Polycarbonate	3
Quartz	4.5
Glass	5
Porcelain	5
Calcite	6.5
Aluminum oxide	9.3
Tantalum pentoxide	26
Niobium pentoxide	42
Mica	58
Epsilan	5000
Ceramic, class 1	>10 … 500
Ceramic, class 2	>500 …>20000

Tab. 1.1: Overview of the permittivity of different dielectrics

By using a dielectric with high relative permittivity (ε_r), a higher capacitance can be achieved with the same package (see Chapter 2.1 Capacitance of a capacitor) or if the same capacitance is required, this can be achieved with a smaller package. This explains the use of different materials as dielectrics in different capacitors and the associated maximum possible capacitance in correlation with the package.

In the historic development of capacitors, alongside the constructive attainment of higher capacitances, it was the search for new suitable dielectrics that made an essential contribution towards attaining higher capacitances and in reducing package sizes.

Every insulator is at the same time a poor electrical conductor. This results in dielectric losses, which have to be especially considered with alternating currents. The occurring losses are caused by the currents in the insulation resistance, in surface resistance and from the polarization effects. The resulting power loss P_V is determined as follows:

$$P_V = 2 * \pi * f * C * U^2 * \tan \delta \qquad (1.6)$$

The material and frequency-dependent dissipation factor tan δ is explained in more detail in Chapter 2.8 Dissipation factor and quality factor Q.

1.7 Displacement flux and displacement flux density

Displacement flux ψ

An electric field originates from the displacement of charges. As long as a capacitor is charging, a current flows – the so-called displacement current. The quantity of charge displaced by the electric field in this process is known as the displacement flux ψ. The displacement flux consists of the sum of all field lines of the electric field. Here the displacement flux is the effect passing through an area and is caused by the charges. Nothing actually moves here, but rather the flux is characterized by a vectorial effect on a unit of charge. The field lines between the plates of a capacitor are therefore the flux lines whose totality characterizes the flux. The displacement flux is directly proportional to the charge that generates the flux. The displacement flux equates to the sum of the magnitudes of all separate charges and the following applies:

$$\psi = Q \qquad \text{with } [\psi] = \text{As} \qquad (1.7)$$

Figure 1.5 illustrates the relationships between the displacement flux and the displacement flux density:

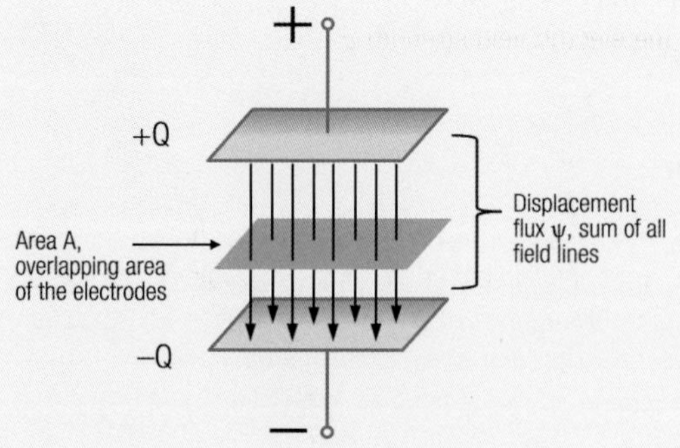

Fig. 1.5: Principle representation of the displacement flux

Dependent upon the field pattern, the area A through which field line pass perpendicularly is of a different size.

Displacement flux density D

The strength of the electric field is described by the displacement flux density *D* (also known in the literature as electric field density). The displacement flux density *D* is the displacement flux per unit area through which field lines pass perpendicularly. The displacement flux density is determined by the quotient of the quantity of charge on the plates, which in this case corresponds to the displacement flux ψ and the effective area A through which field lines pass:

$$D = \frac{\Psi}{A} = \frac{Q}{A} \qquad [D] = \frac{As}{m^2} \qquad (1.8)$$

The displacement flux density is a vectorial quantity and is directly proportional to the existing field strength *E*.

If there is a vacuum between the plates, the flux density can be determined by multiplying the electric field constant and the electric field strength:

$$D = \varepsilon_0 * E \qquad \text{(vacuum as dielectric)} \qquad (1.9)$$

If another insulating material or another dielectric is introduced, the following equation applies for calculating the displacement flux density:

$$D = \varepsilon_0 * \varepsilon_r * E \qquad \text{(arbitrary dielectric)} \qquad (1.10)$$

In case of any other dielectric, the displacement flux density is determined by multiplying the electric field constant ε_0 with the relative permittivity of the respective dielectric ε_r and the electric field strength *E*.

1.8 Capacitor

Capacitor

A capacitor describes every arrangement for storing stationary electric charges. The structure of a capacitor always consists of two conducting surfaces, the so-called electrodes (often termed as sheets). They are always separated from one another by an insulating material, the dielectric. The principle construction of a capacitor is illustrated as follows:

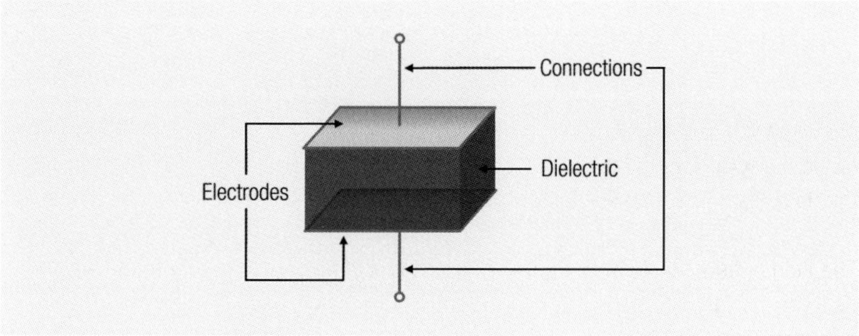

Fig. 1.6: Principle construction of a capacitor

Plate capacitor

The most commonly used arrangement as a technology is the plate capacitor. The principle structure of a plate capacitor consists of two metal plates or foils and a dielectric in the space between them. A plate capacitor and its essential parameters are presented in the following Figure 1.7:

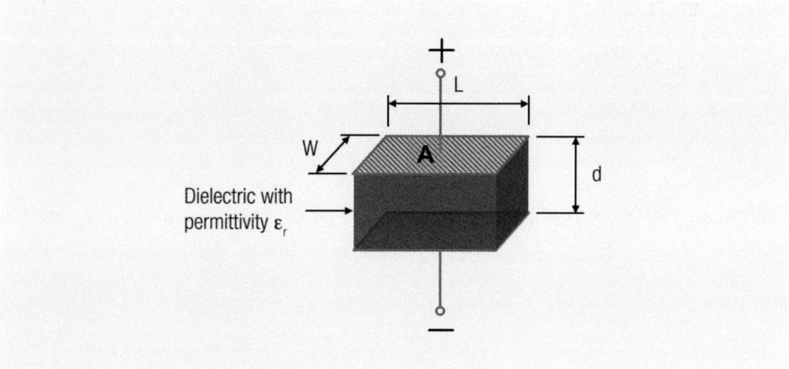

Fig. 1.7: Principle representation of a plate capacitor

Capacitance C

The capacitance C of a capacitor describes its capacity to store electric charges (capacitance is described in more detail in Chapter 2.1 Capacitance of a capacitor). In the case of a plate capacitor, as represented above, its capacitance C can be determined with the following equation:

$$C = \varepsilon * \frac{A}{d} = \varepsilon_0 * \varepsilon_r * \frac{A}{d} \qquad (1.11)$$

The area *A* is determined from the length *L* and width *W* of the electrodes:

$$A = L * W \tag{1.12}$$

The capacitance *C* is calculated from the field constant ε_0, the relative permittivity ε_r of the dielectric used, the effective area *A* (the overlapping area of the electrodes) and the thickness *d* of the dielectric or the separation produced between the electrodes.

The field pattern within an ideal plate capacitor is homogeneous, so that the electric field prevailing in the plate capacitor is of the same magnitude at every point of the field and is uniform throughout. The capacitor types used in modern electronics largely correspond with the principle structure of a plate capacitor.

A capacitor, as an electronic component, is able to store electrical energy and to release it again. The energy release takes place at a defined rate over a certain period depending on its design characteristics.

A capacitor is an energy reservoir, which blocks the direct flow of current with DC voltage and allows the flow of current with AC or pulsating voltage depending on its capacitance and the given frequency. So the capacitor can assume a different role depending on the circuit:

- in a DC circuit it is a charge storage device
- in an AC circuit it is a frequency-dependent resistor

Energy

How much energy a capacitor has stored can be determined from the following formula:

$$E = \frac{C * V^2}{2} = \frac{C * U^2}{2} \tag{1.13}$$

On account of its above properties, the capacitor is an indispensable component in today's electronics. Approximately two-thirds of the passive component market volume is attributable to capacitors. The market for capacitors in relation to sales encompassed a total volume of approximately 13 bn. € in 2012.

Capacitor types

There are various types and constructions of capacitors. Capacitors with fixed capacitance predominate in modern electronics. The main capacitor types of these are presented in the following overview:

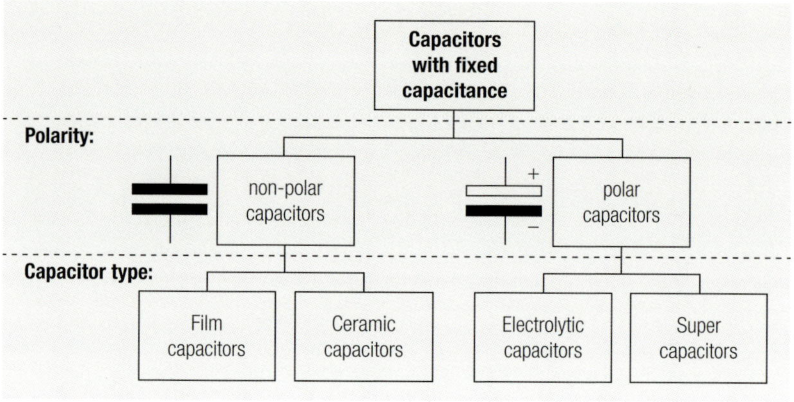

Fig. 1.8: General overview of today's most significant capacitor types with fixed capacitance

Besides the above capacitors with fixed capacitance, there are also capacitors with variable capacitance. Typical examples are rotary or trim capacitors. They play a rather subordinate role in modern electronics as compared with the above capacitors.

The critical distinction is between non-polar and polar variants of capacitors with fixed capacitance. A non-polar capacitor can be operated with DC and AC voltage and it is irrelevant how it is polarized, as the capacitor's electrodes are the same type or are constructed symmetrically.

However, only a DC voltage may be applied to a polar capacitor whose polarity does not change. The electrodes are distinguished as anode and cathode for these designs. The anode has to be connected to the positive and the cathode to the negative potential. The anode and cathode are not the same types for this design. If the polar capacitor is connected incorrectly, damage is caused or there is total failure of the capacitor.

The future development of capacitors includes two essential approaches. On the one hand, the aim is to advance miniaturization in order to keep up with increasing integration density in the electronics industry. On the other hand, the clear focus lies in increasing capacitance, as the capacitor, like the classical rechargeable battery, is a potential energy storage device and with increasing capacitance and advancement it is becoming increasingly attractive.

The important characteristics and parameters of a capacitor are presented and explained in Chapter 2. Subsequently, the various capacitor types are considered in detail in Chapter 3.

Abc of Capacitors

Capacitor Characteristics

Basic principles, characteristics and capacitor types

2 Capacitor Characteristics

The essential characteristics for a capacitor are presented and explained in detail in this chapter. These characteristics are crucial in the selection of a capacitor for a certain application.

2.1 Capacitance of a capacitor

Capacitance C

The most important characteristic of a capacitor is its capacitance C. The capacitance C describes the property of a capacitor's capability to store electrical energy if a (given) voltage U is applied. Capacitance denotes how many units of charge can be stored in the capacitor per voltage unit. Furthermore, the capacitance is important for the AC resistance of a capacitor at certain frequencies, essentially the properties when used in filters.

The unit in which capacitance is specified is a Farad. This was named after the English natural scientist and prominent experimental physicist Michael Faraday (1791 to 1867). A capacitor which is charged to one Volt with one Coulomb of charge (this means that one Ampere of current flows for one second) has a capacitance of one Farad.

The following relationship applies:

$$1\mathrm{F} = 1\frac{\mathrm{AS}}{\mathrm{V}} \qquad (2.1)$$

The capacitance of a capacitor is normally specified in pF, nF or µF. The most common capacitors today lie within this order of magnitude. Values greater than one Farad are attained in the field of super-capacitors. The trend is towards achieving increasingly large capacitance values.

The capacitance of a capacitor essentially depends on the area jointly covered by the electrodes, the separation of the electrodes, the dielectric used and its thickness (see Chapter 1.8 Capacitor).

The capacitance of a capacitor can be increased by means of the following design parameters:

- increase in the effective area of the electrodes
- reduction in the separation between the electrodes
- thinner layer of dielectric
- increase in insulation with a suitable dielectric with higher permittivity or improved dipole formation

2.1.1 Dependence on voltage

By applying a voltage, some insulators used in capacitors as dielectric experience a change in permittivity ε_r and consequently a reduction in capacitance. This can be explained in the case of the dielectrics mentioned here in that once a specific voltage is applied to the electrodes, a defined polarization of the molecules occurs. This reduces the permittivity ε_r of the material or dielectric.

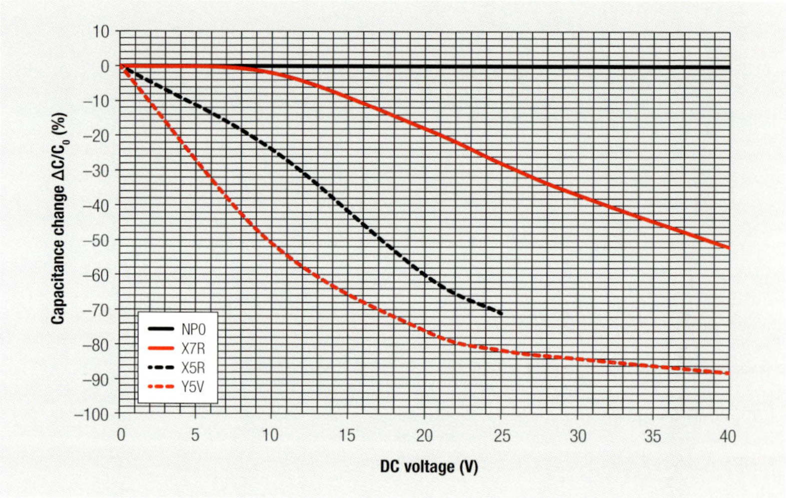

Fig. 2.1: Voltage-dependent capacitance change, example: ceramic capacitors

2.1.2 Dependence on frequency

Capacitor types exist for which a frequency-dependent change in permittivity ε_r occurs. Generally the capacitance of a capacitor drops when the voltage applied increases.

This can be illustrated with the example of an X2 film capacitor in which polypropylene is used as the dielectric. The graph in Figure 2.2 shows the percentage change in capacitance over the frequency band from 1 kHz to 100 kHz.

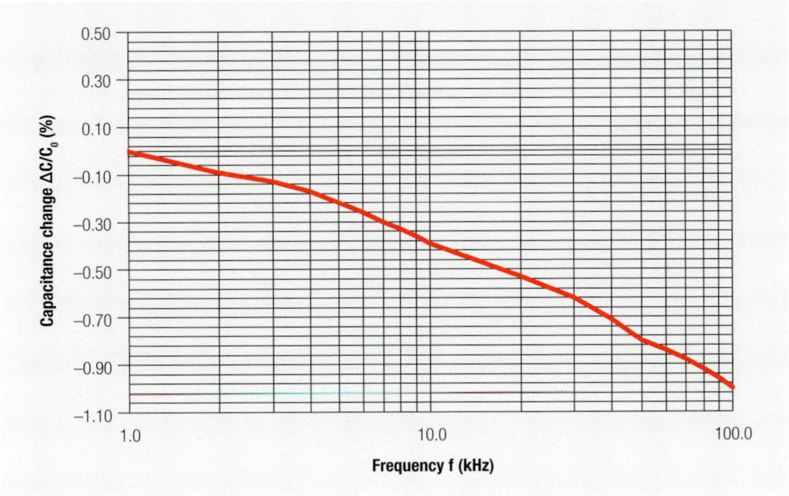

Fig. 2.2: Frequency-dependent capacitance change, example: MKP film capacitor

2.1.3 Dependence on time

Changes in capacitance can occur over time or use in the application. Possible effects:

a) time-limited capacitance change due to temperature change

If the capacitor is subjected to large temperature changes, which are still within the permissible and specified temperature limits, it may occur that the capacitor does not show the intended capacitance value for a certain time The dependence of capacitance on temperature is presented in detail in Chapter 2.10 Temperature coefficient/temperature dependence.

b) Capacitance change due to aging

Aging of a capacitor is caused by its type and the materials it contains. The respective aging behavior has to be taken by the user from the capacitor datasheet or ascertained from the manufacturer.

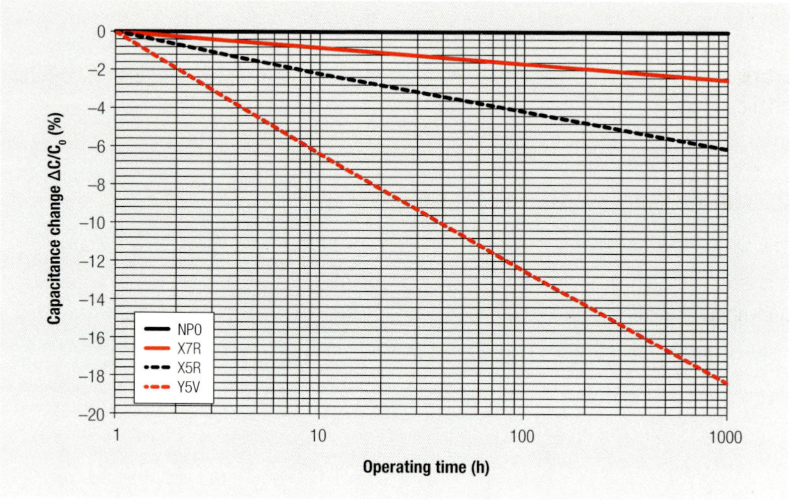

Fig. 2.3: Aging of ceramic capacitors

2.1.4 Capacitance tolerance

Capacitance tolerance

The capacitance tolerance denotes the possible positive and negative deviation from the nominal value of capacitance. The capacitance tolerance is measured and specified by the manufacturer at a standard temperature (usually +20 °C), a standard measurement frequency and under a very low test voltage.

For frequency-determining capacitors, it is e.g. necessary that tight capacitance tolerances are maintained. However, in the case of capacitors used for screening or coupling, the tolerance plays a lesser or insignificant role.

It should be pointed out that the capacitance tolerance is only ever defined as the possible deviation from the capacitance at the time of delivery and does not apply for the entire period of use or the product life of the capacitor.

Capacitors age over their product life and lose capacitance, so the capacitance drops with increasing operating time and the initial tolerance may also be exceeded over time. The aging behavior is also dependent on the dielectric and therefore the capacitance reduces more or less depending on the dielectric.

A low manufacturing tolerance of a capacitor also loses its meaning if the construction or the material properties of the respective capacitor display strong dependencies. This dependency may e.g. be due to temperature, or aging, as described. For instance, Y5V ceramic capacitors are strongly temperature dependent and demonstrate a very different degree of capacitance with fluctuating temperatures.

It should be noted that for applications with precise requirements, capacitors with low dependency behavior (e.g. in respect of temperature) should be selected. Should a capacitor remain in operation for a long period and if the capacitance value must not fall below a certain value for the application to work without faults, the capacitor must be over-dimensioned at the outset. This ensures that the necessary capacitance is also available over time and this value does not drop below the threshold and that the application can be operated without faults.

2.2 Equivalent circuit and its parameters

Equivalent circuit of a non-polar capacitor

The equivalent circuit of a non-polar capacitor and its respective parameters are explained in detail in the following sections. The structure of the equivalent circuit for a non-polar capacitor is as follows:

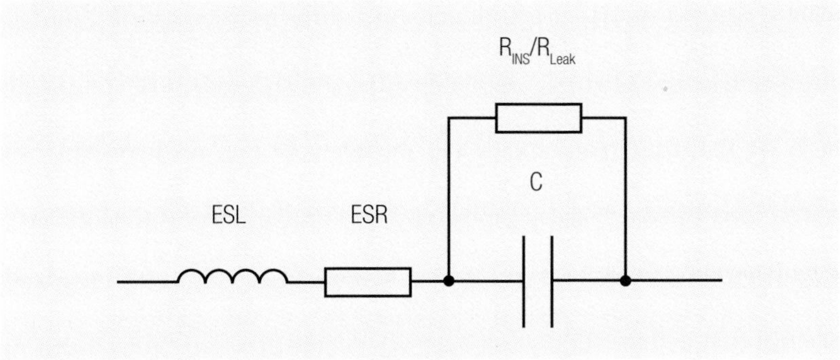

Fig. 2.4: General equivalent circuit of a non-polar capacitor

Insulation resistance R_{INS}
Residual current R_{Leak}

Self-discharge of the capacitor

The equivalent circuit of the capacitor shows the insulation resistance of the dielectric as R_{INS} or in relation to the residual current as R_{Leak}. The capacitor's self-discharge takes place through this resistance. A higher quality dielectric leads to an increase of this resistance and so it becomes apparent that a high quality dielectric increases the possible capacitance, but the actual storage capacity of the capacitor also improves, as the self-discharge drops with increasing R_{INS}. The ohmic components are combined in the equivalent circuit *R* resistance *ESR* and the inductive component is described with the equivalent circuit inductance *ESL*. There is capacitive reactance X_C, inductive reactance X_L and in conjunction with the ESR there is a complex frequency-dependent impedance *Z*.

2.2.1 ESR – Equivalent Series Resistance

Equivalent Series Resistance ESR

The Equivalent Series Resistance *ESR* (also known as R_{ESR}) denotes the effective ohmic resistance existing between the connectors of the capacitor. It combines all ohmic components. These include the connecting wires, contact resistances and the electrodes. Moreover, the *ESR* is a criterion for how well the capacitor passes alternating currents.

The *ESR* can be determined for a specific capacitor (while ignoring the *ESL*) using the following formula:

$$ESR = \frac{\tan\delta}{2 * \pi * f * C} = \tan\delta * X_C \qquad (2.2)$$

The capacitive reactance X_C can be individually determined:

$$X_C = \frac{1}{2 * \pi * f * C} = \frac{1}{\omega * C} \qquad (2.3)$$

With increasing frequency, the reactance X_C decreases.

2.2.2 ESL – Equivalent Series Inductance

Equivalent Series Inductance ESL

Not only inductors exhibit inductance, but in every electrical conductor, due to the flow of current which forms a greater or lesser magnetic field. The Equivalent Series Inductance or L_{ESL} specifies the self-inductance of a capacitor. Depending on the design and construction, the magnetic field is produced to a greater or lesser extent.

The *ESL* is determined by the supply leads, such as wires and other connecting parts and the respective package of the capacitor. As in the case of wound capacitors, a substantial inductance can arise, as each winding connected to the ends of the electrodes, through which current passes, behaves like a coil. Due to their construction, the *ESL* of wound capacitors is highest, but the relevant counter-measures can be adopted here in terms of production technology and construction, such that today's capacitors have a self-inductance on a level with a conductor length equivalent to the length of the capacitor winding.

Moreover, there are film, electrolytic and ceramic capacitors with constructions in which the capacitor has a parallel connection of individual electrodes, whereby there is a very low self-inductance or *ESL*.

The inductive reactance X_L can be determined as follows:

$$X_L = (2 * \pi * f * ESL) = (\omega * ESL) \qquad (2.4)$$

2.2.3 Effective apparent resistance/impedance

Impedance Z

The apparent resistance or impedance *Z* describes how well the capacitor can pass alternating currents at the relevant frequencies. Impedance takes the capacitive and inductive reactance components and the ohmic component *ESR* into consideration. Particularly the parasitic inductive reactance component X_L cannot be ignored here.

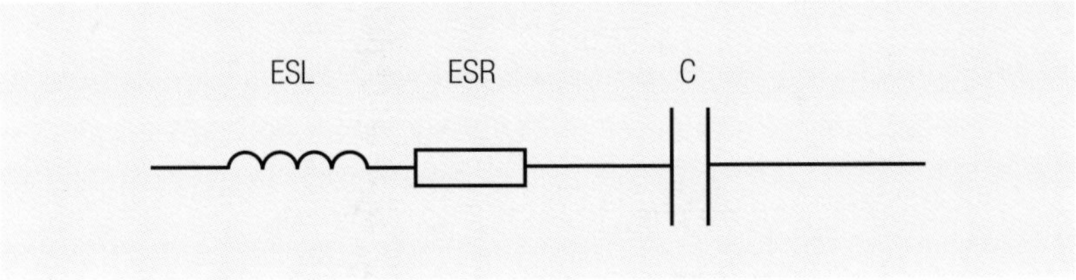

Fig. 2.5: AC equivalent circuit of a non-polar capacitor

Along the lines of the AC equivalent circuit of a capacitor (Figure 2.5) the magnitude of the impedance can be determined from the three components of the equivalent resistance *ESR*, the capacitive reactance X_C and the inductive reactance X_L:

$$Z = \sqrt{ESR^2 + (X_L - X_C)^2} = \sqrt{ESR^2 + \left((2 * \pi * f * ESL) - \left(\frac{1}{2 * \pi * f * C}\right)\right)^2} \quad (2.5)$$

The following Figures 2.6 and 2.7 present the general progression *Z*, the equivalent resistance *ESR*, the capacitive reactance X_C and the inductive reactance X_L as a function of frequency. It shows that all the components are frequency-dependent and how the individual components influence the magnitude of impedance.

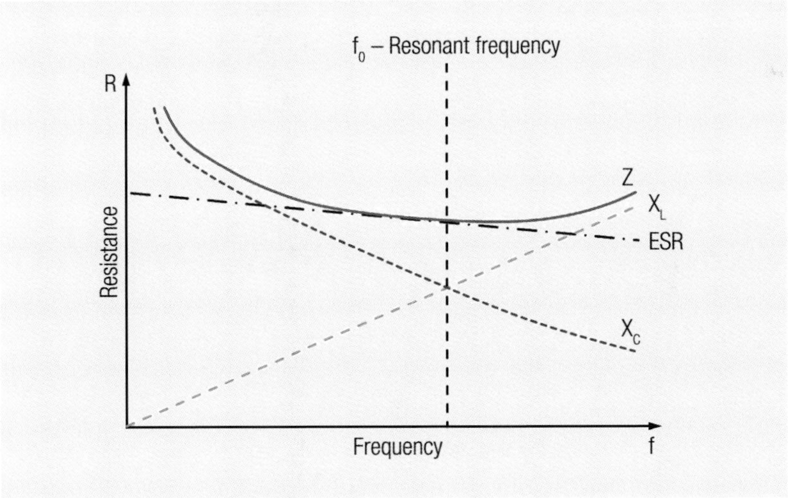

Fig. 2.6: Components of impedance

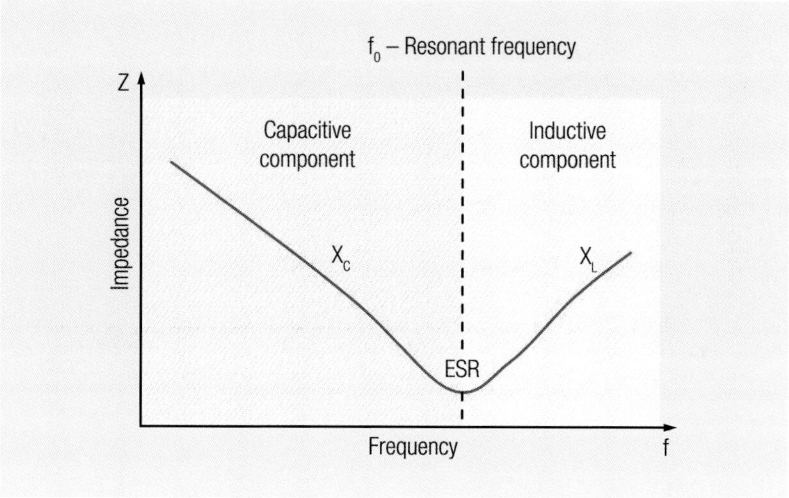

Fig. 2.7: Example of the impedance behavior of a capacitor as a function of frequency

The impedance Z is frequency-dependent and three essential cases (see Figure 2.7) of operating states of the capacitor can be distinguished:

1) $f < f_0$ – below the resonant frequency
→ X_C has the main effect
2) $f = f_0$ – at the resonant frequency
→ only *ESR* takes effect, as X_C and X_L cancel each other
3) $f > f_0$ – above the resonant frequency
→ X_L has the main effect

Even a small *ESL* of the capacitor, which arises from its construction, causes an increase in resonant frequency. If a capacitor is operated at high frequencies, the inductive component X must not be ignored. Operating a capacitor above the resonant frequency, it already acts as an inductance. In this case, attention must be paid to using a capacitor with the smallest possible *ESL*, which has a low inductance construction.

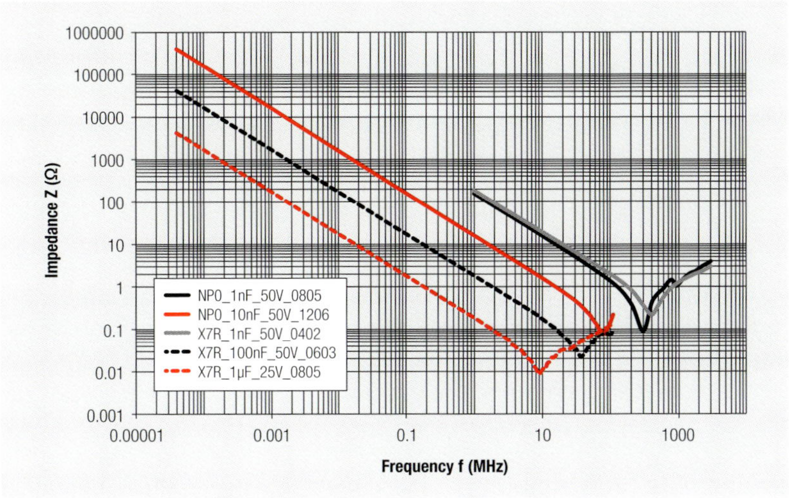

Fig. 2.8: Example: Impedance behavior of ceramic capacitors

2.2.4 Resonant frequency/self-resonance

For applications in which capacitors are used, it is important to consider that the capacitor's supply leads act as inductances and this creates a series resonant circuit of *ESL* and *C* (see configuration in Figure 2.4). As described in section 2.2.2 ESL – Equivalent Series Inductance, this effect can be minimized by designing film capacitors by contacting/soldering connection pins on the front face to the electrode (winding) of the capacitor.

Resonant frequency f_0

The resonant frequency f_0 of the series resonant circuit of *ESL* and *C* can be determined as follows:

$$f_0 = \frac{1}{(2 * \pi * \sqrt{ESL * C})} \quad (2.6)$$

On natural resonance of a capacitor, the impedance *Z* is at its minimum at this frequency, as it then only consists of the ohmic component *ESR* (see 2.2.3 Effective apparent resistance/impedance and Figure 2.7).

2.3 Rated voltage

Rated voltage U_R

As a result of its construction, a capacitor can be loaded up to a defined rated voltage U_R (also known as nominal voltage/operating voltage/working voltage). The rated voltage denotes with which voltage the capacitor may be continuously loaded without it being damaged. The rated voltage of a capacitor is generally distinguished between rated DC voltage and rated AC voltage. These are explained in more detail in the following two sections.

2.3.1 Rated DC voltage

Rated DC voltage U_{RDC}

The rated DC voltage U_{RDC} is the maximum voltage which may be applied continuously to the capacitor. The rated DC voltage specified in the datasheet applies from the minimum permissible temperature limit up to the rated temperature of the capacitor. The specification for the rated DC voltage takes the superposition of DC/AC/pulse voltages into consideration.

2.3.2 Rated AC voltage

Rated AC voltage U_{RAC}

The rated AC voltage U_{RAC} denotes the maximum root-mean-square (RMS) of the AC voltage which can be applied continuously to a capacitor at a specific frequency (e.g. 50 Hz). As with the rated DC voltage, this specification applies from the minimum permissible temperature limit specified in the datasheet up to the rated temperature of the capacitor. The rated AC current is only specified for non-polar capacitors.

2.4 Maximum continuous voltage

Maximum continuous voltage

The maximum continuous voltage (Category Voltage U_C) denotes the maximum permissible voltage value with which the capacitor can be continuously loaded if the capacitor is operated at the maximum temperature limit specified in the datasheet. The specification for the maximum continuous voltage already takes the superposition of DC/AC/pulse voltages into consideration. The maximum continuous voltage is lower than the rated AC voltage and so it is clear that a rise in temperature above the rated temperature (usually +20 °C or +25 °C) requires successive voltage derating up to the maximum specified temperature limit.

2.5 Insulation resistance

Insulation resistance R_{INS}

The insulation resistance R_{INS} denotes the ohmic resistance between the electrodes of a capacitor. It is known as R_{INS}, *IR* or *R*. The insulation resistance is a quality criterion of the capacitor and is determined by the relationship between the DC voltage applied and the insulation current. Usually the measurement is made one minute after the capacitor has charged.

In capacitor datasheets the insulation resistance is either specified with the resistance value [MΩ] or with the time constant τ [s]. In addition there are also standardized R_{INS} specifications in MΩ*µF. With this kind of specification the insulation resistance R_{INS} equals the time constant τ:

$$\tau[s] = R_{INS}\,[M\Omega * \mu F] = R_{INS}\,[M\Omega] * C[\mu F] \qquad (2.7)$$

Insulation resistance is temperature-dependent. If the temperature increases, the insulation resistance reduces. The use of different dielectrics leads to different levels of insulation resistances as a result of their material properties. Dielectrics exist with very high insulation resistances. These are used e.g. in ceramic capacitors (class 1). For such capacitors with very high insulation resistance, it can be ignored in most cases. In such a case, the insulation resistance corresponds to the surface resistance of the capacitor outer shell.

2.6 Residual current (DC leakage current)

Residual current I_{Leak}

The term residual current denotes the DC current that flows through the capacitor when a voltage is applied. The residual current is also known as DC leakage, insulation or creep current. The magnitude of the residual current depends on the conditions at the capacitor, i.e. capacitance, voltage, temperature and time or the time after voltage is applied.

The residual current is specified in the literature with the symbol I_{Leak}, *DCL*, I_{kr} or I_L. In datasheets the residual current is given in nA or µA. There are also standardized *I* specifications with the dimension nA/µFV.

The residual current $I_{Leak\ App}$ for a particular application can be determined as follows:

$$I_{Leak\ App}\ [nA] = I_{Leak}\ [nA/FVC] * C * U_C \quad (2.8)$$

U_C is the maximum working voltage or the peak voltage applied to the capacitor. When calculating the residual current for a special application, the calculation is dependent on the respective capacitance of the capacitor and the voltage U applied to the capacitor.

In practice, if I_{Leak} is not specified, rules of thumb are used to estimate the residual current. Two common rules of thumb used for its determination, which are essentially used for electrolytic capacitors, are the following:

$$I_{Leak\ App}\ [\mu A] = 4 + (0.006 * C[\mu F] * U[V]) \quad (2.9)$$

$$I_{Leak\ App}\ [\mu A] = 0.01 * C[\mu F] * U[V] \quad (2.10)$$

In case of non-polar capacitors, the insulation resistance R_{INS} is usually specified. In contrast, for electrolytic capacitors the residual is usually specified as current I_{Leak}. A high insulation resistance corresponds to a low residual current and vice-versa. If the temperature increases, the insulation resistance of the capacitor decreases.

The residual current occurring in a real application essentially depends on the working voltage and the respective ambient temperature.

It should be pointed out that the residual current of an electrolytic capacitor changes over time. The residual current is highest after direct connection of the voltage and then drops to an operating current in the µA range.

2.7 Self-discharge

Self-discharge

A charged capacitor discharges to a certain degree over time on its own. There is self-discharge via R_{INS}. The discharge occurring is caused by the factors insulation, circuitry, the electrodes and the dielectric itself. Especially if AC voltage is used, there are high losses due to constant repolarization. There are capacitors used for AC voltage applications whose design is better suited for the purpose of minimizing losses.

2.8 Dissipation factor and quality factor Q

Phase shift

An ideal capacitor (purely reactance X_C) has a phase shift, whereby the current is a phase angle of exactly 90° ahead of the voltage. In case of a real capacitor, there are inductive and ohmic resistance components through which the phase shift is reduced accordingly and the result is a phase angle $\phi < 90°$. The difference from the ideal phase angle of 90° is specified as δ.

Loss angle δ

Dissipation factor tan δ

The dissipation factor tan δ (or dissipation factor – DF) denotes the level of the losses for the respective capacitor. Furthermore, the dissipation factor tan δ describes the frequency-dependent relationship of the ohmic resistance component ESR to the capacitive reactance X_C. tan δ is calculated as follows:

$$\tan\delta = DF = \frac{ESR}{X_C} = ESR * 2 * \pi * f * C \qquad (2.11)$$

The dissipation factor tan δ also describes the relationship of the components of (loss-free) capacitive current and the (lossy) ohmic current component which flow at the same time when operating the capacitor. A capacitor with a low dissipation factor thus has lower losses.

Active power P
Reactance Q

The dissipation factor represents the relationship between the reactive power P and the reactance Q.

$$\tan\delta = \frac{P}{Q} \qquad (2.12)$$

It is important to note that the dissipation factor is dependent on temperature and frequency. This is clearly illustrated with the example of an MKP film capacitor.

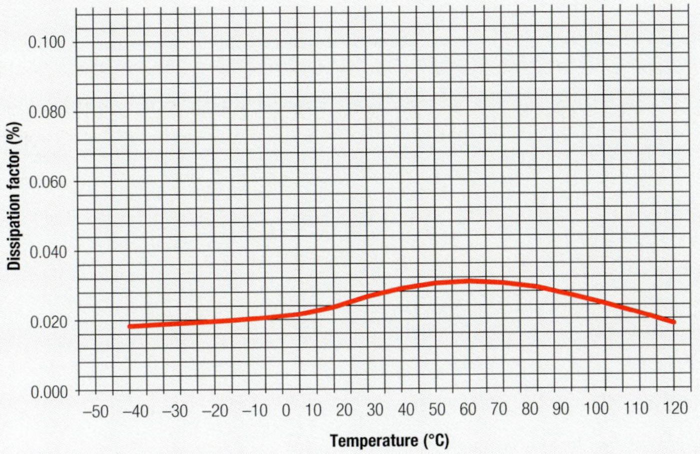

Fig. 2.9: Temperature-dependent dissipation factor, example of MKP film capacitor

The dissipation factor is especially important with AC voltage capacitors loaded with high currents. The loss component generates heat loss depending on the dissipation factor and leads to self-heating of the capacitor. If the dissipation factor tan δ is too large, this can lead to excessive heating of the capacitor and this stress can reduce the effective service life of the capacitor.

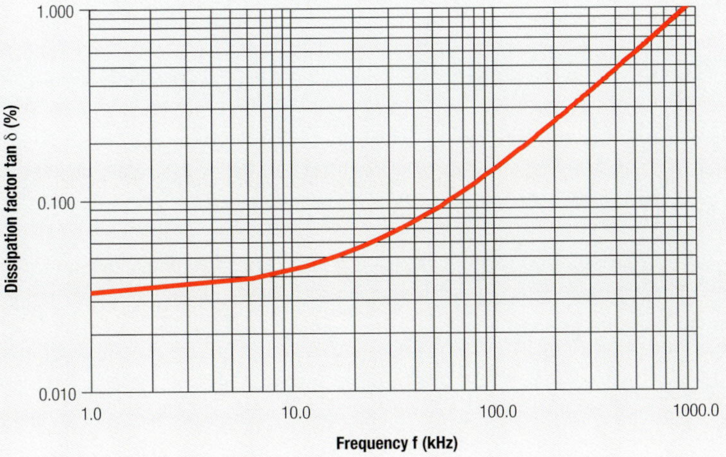

Fig. 2.10: Example of an MKP film capacitor

Quality factor Q_c

The quality factor Q_C of a capacitor can be determined as the reciprocal of the dissipation factor:

$$Q_C = \frac{1}{\tan \delta} = \frac{X_C}{ESR} \tag{2.13}$$

The quality factor denotes the relationship between the energy stored in the capacitor and the energy losses occurring. Like the dissipation factor, it is calculated from the capacitive reactance *X* and the equivalent series resistance *ESR*. The higher the quality factor, the lower the loss of the respective capacitor.

2.9 Dielectric strength/voltage

Dielectric strength/voltage

The dielectric strength/voltage specifies the maximum permissible voltage which may be applied to the capacitor without a flashover occurring between the electrodes. If this voltage level is exceeded, the dielectric no longer insulates adequately. There is a flashover through the dielectric and damage to the capacitor.

The dielectric strength of a capacitor depends on the respective dielectric used and its characteristics. In order to avoid damage to or destruction of the capacitor, it is necessary to ensure that the voltage permissible for the dielectric is not exceeded.

2.10 Temperature coefficient/temperature dependence

Temperature dependence

Capacitors are subject to temperature-dependence. If the temperature of a capacitor changes (e.g. due to increasing ambient temperature or operation of the application), the capacitance also changes. The capacitance change is explained in that the permittivity ε_r of a dielectric is temperature-dependent. The change in capacitance with changing temperature can be both positive and negative. It is essentially dependent on the capacitor's construction and the materials used (predominantly the dielectric).

Temperature coefficient TC

The temperature-dependence of a capacitor is described with the temperature coefficient *TC* (also known as *TCC* or α_C in the literature). *TC* is usually expressed in ppm/°C or in %/°C.

In the ideal case, no change in capacitance occurs with changing temperature. There are types of capacitor that have a temperature coefficient $TC = 0$ (e.g. COG or NPO ceramic capacitors) and there is no change in capacitance due to temperature changes or fluctuations. This is necessary, because some applications need a precisely calculated capacitance and otherwise their proper function could not be guaranteed for fluctuating or differences in temperatures.

There are also capacitor types with highly pronounced temperature dependence. In such cases, the change in capacitance is specified over the entire temperature range.

The temperature coefficient *TC* in ppm/°C is derived as follows:

$$TC = \frac{1}{C_1} * \frac{C_2 - C_1}{\vartheta_2 - \vartheta_1} * 10^{-6} = \frac{1}{C_1} * \frac{\Delta C}{\Delta\vartheta} * 10^{-6} \quad (2.14)$$

TC ... temperature coefficient [ppm/°C]
C_1 ... capacitance of the capacitor at +20 °C
C_2 ... capacitance at the prevailing temperature ϑ_2
ϑ_1 ... reference temperature of +20 °C
ϑ_2 ... current temperature of the capacitor

If the temperature coefficient TC is known, the following parameters of a capacitor can be determined:

1) resulting difference in capacitance with change in temperature:

$$\Delta C = TC * C_1 * \Delta\vartheta * 10^{-6} \quad (2.15)$$

2) resulting percentage difference in capacitance with change in temperature:

$$\Delta C[\%] = \frac{C_2 - C_1}{C_1} * 100\% = TC * \Delta\vartheta * 10^{-4}\% \quad (2.16)$$

3) resulting capacitance with a known change in temperature:

$$C_2 = C_1 * (1 + TC * \Delta\vartheta * 10^{-6}) \quad (2.17)$$

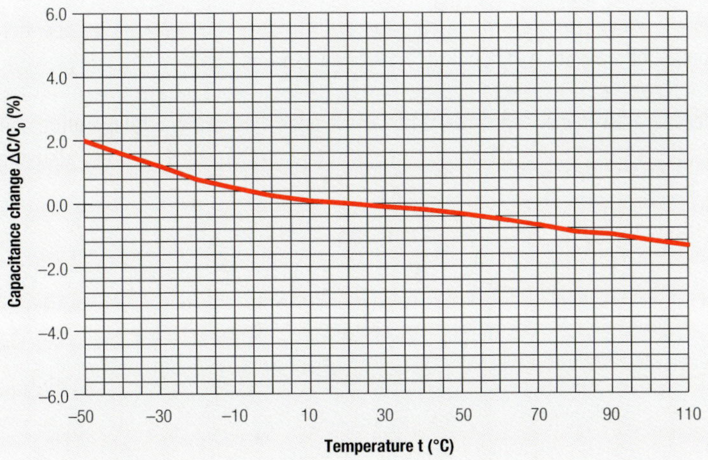

Fig. 2.11: Temperature-dependent change in capacitance, example of MKP film capacitor

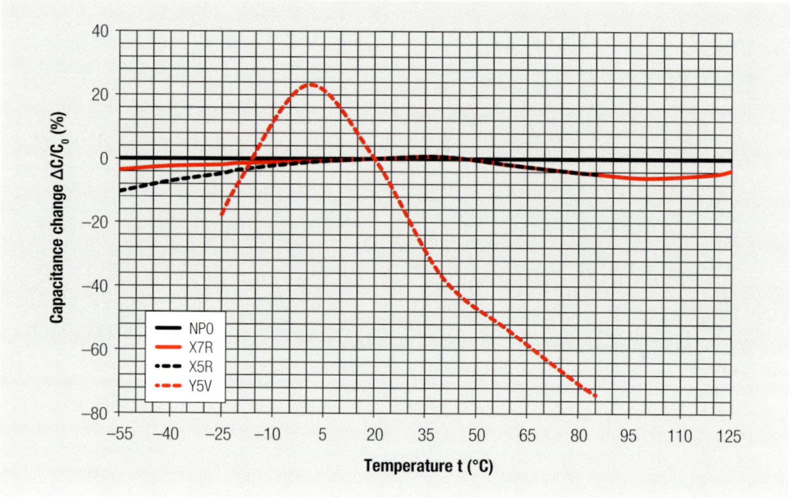

Fig. 2.12: Temperature-dependent change in capacitance, example of ceramic capacitors

2.11 Pulse rise time

Pulse rise time dV/dt

The pulse rise time *dV/dt* (also known as edge steepness or permissible speed of voltage change) denotes with what changes in voltage over a certain time the capacitor may be loaded. It therefore describes the maximum permissible rate of change of voltage with time. The pulse rise time (rate of voltage change) is termed *dV/dt* or $\Delta U/\Delta t$ in the literature.

The permissible pulse rise time is always expressed in voltage per time and this is usually in the unit V/µs.

Statement of the permissible pulse rise time can be illustrated with an example. If, according to the datasheet, a capacitor has a permissible pulse rise time of 200 V/µs, the following changes are conceivable, for example:

- 200 V voltage change/swing over a period of 1 µs
- 100 V voltage change/swing over a period of 0.5 µs
- 400 V voltage change/swing over a period of 2 µs

The example shows that the essential statement here is the rate of change of voltage over a defined period. The above examples of voltage changes can be extended at will, as long as the ratio of voltage change and time do not exceed 200 V/µs. It is important to note, however, that the maximum permissible voltage increase/swing *dV* or ΔU for which the pulse rise time in the datasheet applies, corresponds with the rated DC voltage of the capacitor and this must not be exceeded.

The permissible rate of voltage change is always in relation to current and capacitance and the following relationship applies:

$$\frac{\Delta U}{\Delta t} = \frac{dV}{dt} = \frac{I}{C} \tag{2.18}$$

If the operating voltage from the application is lower than the rated voltage of the capacitor, higher pulse rise time are possible. In order to be in a position to determine these, the pulse rise time dV/dt_{spec} ($\Delta U/\Delta t_{spec}$), the rated DV voltage U_R and the voltage swing of the application ΔU_{app} have to be known. The pulse rise time permissible for the application is calculated as follows:

$$\left(\frac{\Delta U}{\Delta t}\right)_{app} = \left(\frac{\Delta U}{\Delta t}\right)_{spec} * \frac{U_R}{\Delta U_{app}} \tag{2.19}$$

2.12 Pulse loading capacity

Pulse loading capacity

The pulse loading capacity (pulse stressing) of a capacitor is always in relation to the permissible rate of voltage change. If there is a sudden change in voltage dV in a time period dt, there is a resulting current I_C at the capacitor.

Here the current I is the product of the rated capacitance C and the permissible rate of voltage change and the pulse rise time dV/dt:

$$I = C * \frac{dV}{dt} \tag{2.20}$$

If the permissible dV/dt of the capacitor is exceeded, this leads to an excessive current and hence damage to the capacitor, e.g. contacting of the connecting wires with the electrodes.

2.13 Dielectric absorption

The effect can be observed in capacitors following complete discharge of the capacitor after several seconds up to minutes in duration, in that a small voltage can be measured at the capacitor's connectors. For some HV capacitors, this effect can appear after hours or even days.

Dielectric absorption

The recharging effect occurring due to dielectric absorption is caused by polarization processes of the dipoles of the dielectric used.

The degree of the recharging that occurs is mainly dependent on the capacitance of the capacitor and the type and thickness of the dielectric used.

In practice, this means that after high voltage testing on insulated capacitors, a very high voltage can occur that can destroy equipment during repeated measurements. A noticeable electric shock for the user is also possible. From case to case, capacitors should therefore be short-circuited after HV DC testing.

2.14 Ripple current

Ripple current

In a capacitor application, the ripple current is a measurable AC voltage at the capacitor caused by a parasitic AC voltage (regardless of what form or frequency) applied to a DC voltage channel.

If a capacitor is affected by a ripple current, the existing AC voltage causes charging and discharging processes to occur in the capacitor which are superposed on the DC voltage. This results in an effective current through the capacitor's equivalent series resistance *ESR*. The power loss is converted to heat and leads to self-heating of the capacitor. Ripple current mainly occurs after rectification of an AC voltage.

Ripple current is generally not harmful for non-polar capacitors if the specified electrical parameters and temperature limits are maintained.

Ripple current is especially critical for polar capacitors or electrolytic capacitors. The maximum permissible ripple current is therefore always specified for electrolytic capacitors. The maximum permissible ripple current of an electrolytic capacitor denotes the ripple current the capacitor can be loaded to without damage and such that it still complies with the service life forecast. If an electrolytic capacitor is loaded with too much ripple current, the inside temperature of the capacitor rises beyond the permissible level. This can damage the capacitor and lead to its failure. This situation is also a reason why very low *ESR* values are aspired for with electrolytic capacitor designs and why development is aimed at new designs, as well as increasing capacitance.

2.15 Complex permittivity

Operating a dielectric with AC voltage, the relative permittivity ε_r of the dielectric depends on which frequency the capacitor is operated. This can be explained in that the polarization mechanisms taking place in the dielectric are frequency dependent and they influence the relative permittivity ε_r of a dielectric.

Up to three different polarization mechanisms may be responsible for the polarization, depending on the material and dielectric. These polarization mechanisms are:

Electronic polarization
The dielectric is influenced by an applied electric field.
The electric field causes a slight shift of electrons within the atomic shell to the side of the atom towards the positive electrode. The protons, on the other hand, tend to shift towards the negative electrode. If the electric field changes its direction over time (AC voltage), the dipoles oscillate back and forth and the electrons and protons align as

described above. The masses shifted by this polarization mechanism are very low and therefore electronic polarization can take place very quickly. This polarization mechanism can be observed in all materials and occurs at a frequency of approx. 10^{15} Hz (frequency of visible light).

Electronic polarization

Ionic polarization

Ionic polarization

Ionic polarization describes the elastic shift of ions in the ionic lattice. As the mass of ions is significantly higher than the mass of electrons, polarization takes place slower in ionic polarization than with electronic polarization. A frequency of up to 10^{13} Hz is possible for ionic polarization. Ionic polarization disappears above this frequency (and with it, it's component in the relative permittivity ε_r).

Orientation polarization

Orientation polarization

Orientation polarization describes the alignment of permanent dipoles. Without an electric field applied, the dipoles in the dielectric are distributed statically. Once an electric field is applied, there is alignment or orientation of the dipoles along the electric field. Precise alignment of the dipoles parallel with the field direction is not achieved due to the prevalent thermal movement of the molecules.
As a result of the mass of the dipoles aligned in this polarization mechanism, this proceeds relatively slowly. In addition, this mechanism is temperature-dependent, as a temperature increase counteracts the alignment or polarization of the dipoles. Orientation polarization disappears at frequencies above 10^9 Hz.

The polarization mechanisms described above are illustrated in the following figure:

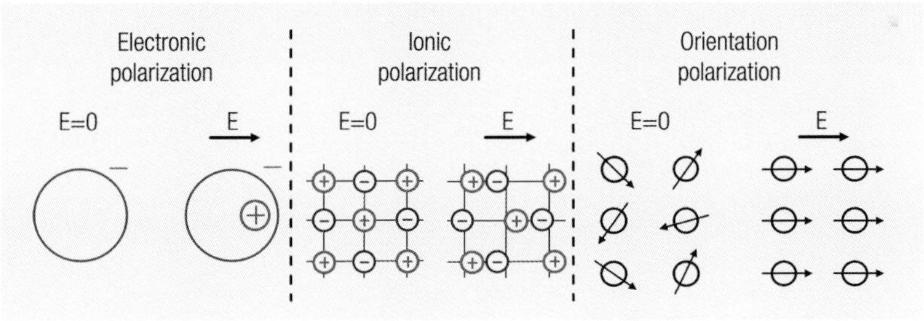

Fig. 2.13: Mechanisms of polarization of dielectrics

The relative permittivity ε_r consists of the three polarization components described. As the respective polarization effects are frequency dependent, the resulting relative permittivity ε_r also shows a corresponding dependence on frequency. Figure 2.14 illustrates the general progression of relative permittivity ε_r as a function of frequency:

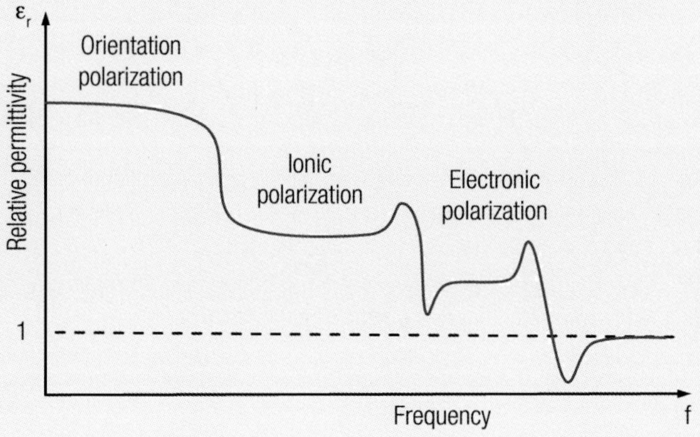

Fig. 2.14: General progression of the relative permittivity as a function of frequency

As shown in the figure, orientation polarization only disappears with increasing frequency. Increasing the frequency further, ionic polarization disappears. If very high frequencies are applied, it may occur that electronic polarization is so far delayed that the excitation is in phase opposition. This means that the relative permittivity ε_r drops to a value below 1, before the electronic polarization disappears completely with further increase in frequency.

In AC operation, two effects occur in the dielectric that lead to electrical losses. These effects are:

- Residual conductivity
- Orientation polarization

Every insulator and therefore every dielectric always has a certain residual conductivity. Residual conductivity may also be observed in DC operation of a capacitor.

Orientation polarization only occurs in AC operation, however. In orientation polarization of polar molecules, the dielectric heats up, as the constantly repolarized molecules or dipoles transmit oscillations or shocks to the neighboring/surrounding molecules and this produces friction. The losses occurring are particularly high at resonance. This is the case if the frequency of the AC voltage coincides with the natural frequency of the molecules. Residual conductivity and orientation polarization act as active power and can be described with a resistance R or conductance G connected in parallel with the ideal capacitor.

The resulting complex total conductance is calculated as follows:

$$Y_{ges} = G + (j * \omega * C) = G + (j * 2 * \pi * f * C) \qquad (2.21)$$

Complex relative permittivity ε_r

The complex relative permittivity ε_r of a lossy capacitor is defined as follows:

$$\varepsilon_r = \varepsilon_r' - j * \varepsilon_r'' \quad (2.22)$$

The real part ε_r' corresponds to the relative permittivity of a specific material or dielectric. The imaginary part ε_r'' describes the dielectric losses. The dielectric losses denote the level of conversion of electric energy into heat energy.

The dielectric dissipation factor can be determined as follows:

$$\tan \delta = \frac{\varepsilon_r''}{\varepsilon_r'} \quad (2.23)$$

The consequence for a capacitor is that in AC voltage operation at high frequency it should be checked whether the required capacitance is still available from the dielectric used. At high frequencies it may be the case that the disappearance of polarization effects and the associated drop in the relative permittivity leads to a reduction in capacitance (as ε_r drops with increasing frequency specifically for a dielectric). The losses occurring at the capacitor therefore increase.

Abc of Capacitors

Capacitor Types

Basic principles, characteristics and capacitor types

3 Capacitor Types

As mentioned already, there is a wide range of different capacitors. The capacitors briefly introduced in Chapter 1.8, which play the most important role in industry are presented in detail in the following chapters and in addition other existing designs are shown.

The following chapters serve to provide an essential understanding of the existing capacitor designs and to introduce and describe their basic features. Each of the capacitor types could be dedicated its own reference book.

3.1 Overview of existing capacitor types

Capacitor types

The capacitor types can be broadly divided into those with fixed capacitance and those with variable capacitance. The most important are the fixed capacitance capacitors, but capacitors with variable capacitance also exist. These include rotary or trimmer capacitors.

Capacitors with fixed capacitance can be divided into film, electrolytic, ceramic and super conductors. The film, electrolytic and ceramic capacitors are described in more detail in the following chapters and are the capacitors most commonly found in electronics.

All existing capacitor types and their classification are presented in the following overview:

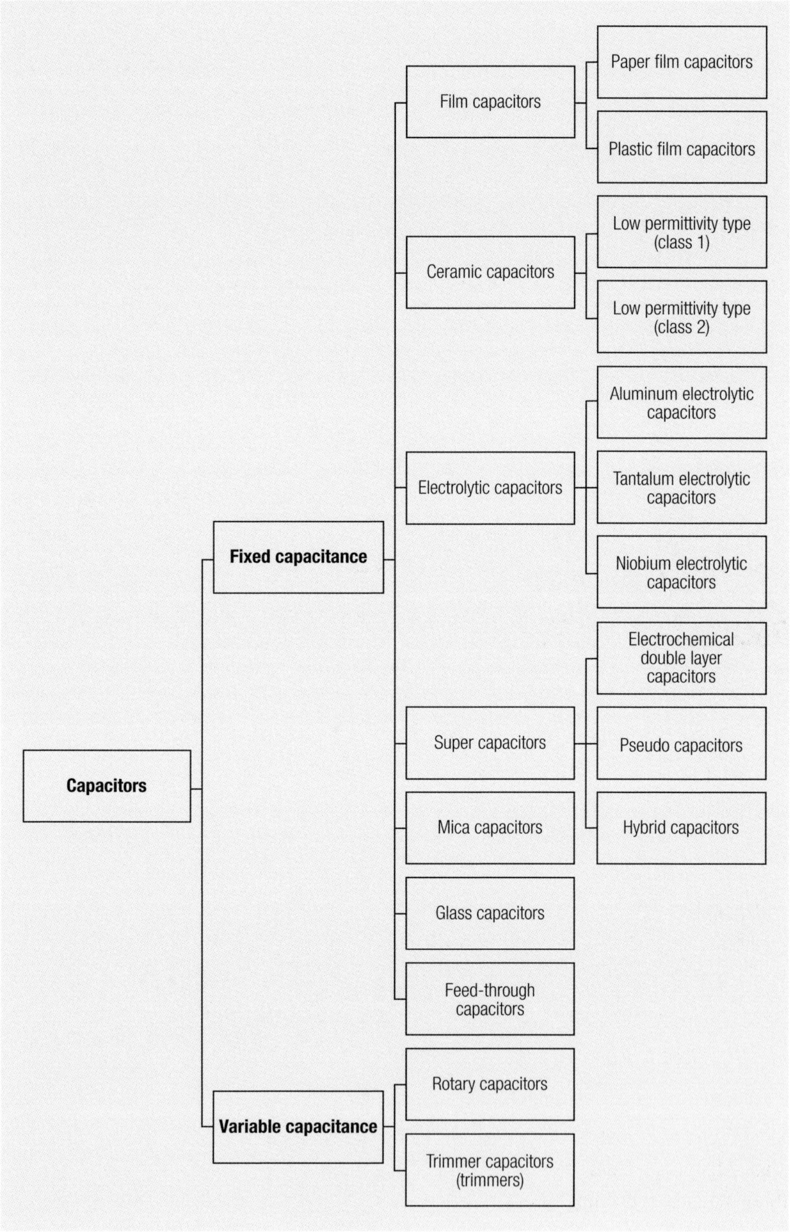

Fig. 3.1: Overview of existing capacitor types

3.2 Film capacitors

Film capacitors

On account of their construction and the materials used, these capacitors are known as film capacitors. Film capacitors have no polarity and have a fixed and defined capacitance.

To construct the capacitor, either two metal films with an intermediate film, or films of dielectric on which a metal layer is vapor deposited, are wound. This design is therefore also known as a wound capacitor.

Film capacitors either use paper or plastic films as dielectric. As described, the electrodes are either metal films (in general aluminum) or aluminum layers vapor deposited or metalized directly on the dielectric used. The electrodes are contacted with a metal layer on the end face of the capacitor (schoopage = metal flame spraying). The component connectors are attached via this contact surface.

As a result of their different construction, the film capacitors can be distinguished as follows:

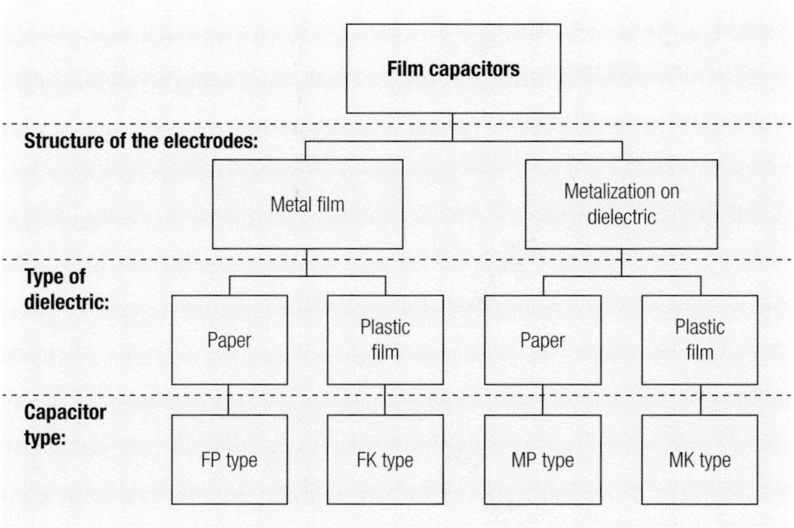

Fig. 3.2: Overview of film capacitor types

The FP and PM types are metal-paper capacitors, whereas the FK and MK types are plastic film capacitors.
(M = metalized; F = film; K = plastic; P = paper)

Paper has several disadvantages in comparison with plastic as a dielectric:

- high moisture uptake
- mechanical shrinkage
- low relative permittivity ε_r

Owing to the low relative permittivity of paper, the same capacitance and voltage stability values can be obtained with plastic films of a smaller package size or higher capacitance values with the same package size.

There are two essential construction types of plastic film capacitors (FK and MP), which differ in the type of capacitor electrode.

FK capacitors

FK capacitors

Individual metal films are wound with an intermediate plastic film, which serves as a dielectric. So the electrodes and the dielectric are each separate films. Through subsequent thermal treatment, the plastic films are shrunk after winding to mechanically stabilize the winding and to raise the moisture protection between the individual films.

MK capacitors

MK capacitors

A plastic film which serves as the dielectric is vapor deposited with a thin metal layer in this design. Here the electrodes are vapor deposited on the dielectric directly. The thickness of the vapor deposition layer is usually in the range 10–50 nm.

The difference in design between MK and FK capacitors leads to a difference in the construction and the characteristics of the capacitor. MK capacitors display the following differences to FK capacitors:

- lower permissible voltages and currents
- smaller package size achievable
- higher dissipation factor tan δ
- lower insulation resistance R
- self-healing properties

Typical plastics used as dielectrics with film capacitors include:

- polycarbonate (Makrofol)
- polyphenylene sulfide
- polyester (Mylar)
- polypropylene
- polystyrene (Styroflex)
- polytetrafluorethylene (Teflon)
- cellulose acetate

Depending on the type of film and dielectric used, the following IEC 60062 codes are used for film capacitors. This is presented for some typical capacitors in the following table:

Dielectric	Code for the FK capacitor	Code for the MK capacitor
Polyester (PETP)	KT	MKT
Polycarbonate (PC)	KC	MKC
Polypropylene (PP)	KP	MKP
Polystyrene (PS)	KS	MKS

Tab. 3.1: Coding for typical film capacitors

The use of coding can be illustrated with an example. For example, the code MKP identifies that this capacitor uses a polypropylene film as the dielectric and the electrodes are metalized directly on the polypropylene films. If this were now a KP version, then there would be a polypropylene film as dielectric, but the electrodes are separate films and are not metalized on the plastic film.

The properties of a film capacitor are essentially influenced through the film material used and the design of the electrodes, as this material leads to different electrical properties.

The following figure shows the schematic structure of a plastic film capacitor (MK type):

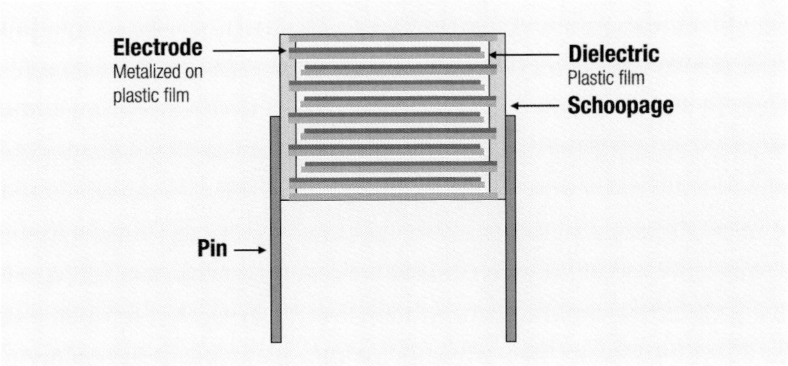

Fig. 3.3: Schematic structure of an MK film capacitor

The inner structure of the MK film capacitor shows that two metalized films are wound offset relative to each other. These are contacted collaterally with so-called schoopage. Schoopage is a metalization layer on the end face of the capacitor and contacts the individual layers of a film metalization or an electrode. The connection to the capacitor is achieved via the connection pins which are attached by schoopage.

The winding shown in the figure is usually encapsulated. For example, the winding is put into a plastic box and is sealed air-tight with a potting compound.

Fig. 3.4: Example of MKP capacitors

There are various possibilities of arranging the films, e.g. inner series connection of the films. Reinforced films and double-sided layers are also encountered. The voltage stability/pulse stability can be increased with this change in arrangement of these versions. Versions also exist for which films are not wound, but individual film layers are stacked to form a package. This is known as a stacked film capacitor.

3.3 Electrolytic capacitors

Electrolytic capacitors

Unlike other capacitor types, the name of a electrolytic capacitor does not allow conclusions to be drawn as to its actual dielectric. On account of the polar or non-symmetrical structure of an electrolytic capacitor, the electrodes can be distinguished as anode and cathode. In the case of an electrolytic capacitor, the anode consists of a metal film (usually aluminum) and the cathode is a conductive electrolyte. The electrolytic capacitor therefore only has one metallic electrode in contrast with other capacitors (in this case the anode).

An oxide layer on the metal film of the anode acts as the dielectric between the two electrodes (anode and cathode). This oxide layer insulates the electrodes from one another.

Electrolytic capacitors may be distinguished on the basis of two essential criteria. These are the electrode material used and the property of the electrolytes. Possible configurations are presented as follows:

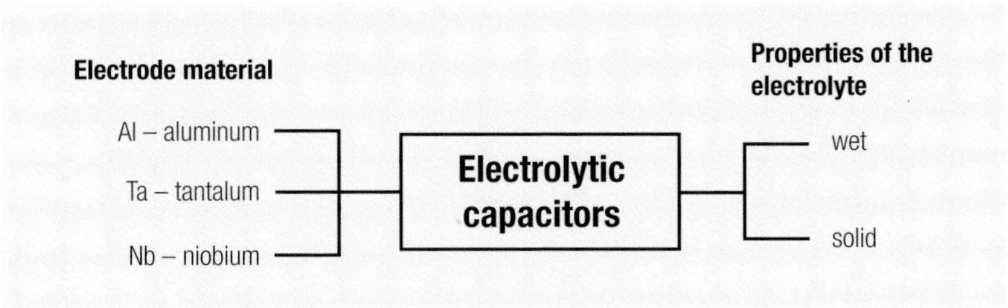

Fig. 3.5: Possible structure of an electrolytic capacitor

In today's electrolytes the electrodes consist of aluminum, tantalum or niobium and there are designs with wet (gelatinous) electrolyte or solid electrolyte. There are also hybrid solutions in which a wet/solid electrolyte version is used.

The design of the electrolytic capacitor has a major influence on its electrical parameters (capacitance, ESR, residual current, etc.) and on the volume necessary to attain a required capacitance.

Depending on which electrode material is used for the anode, the dielectric of the electrolytic capacitor consists of an oxide of the anode material. The respective oxide layers have different relative permittivities:

Anode electrode material	Dielectric	Relative permittivity (ε_r) at +20 °C
Al – aluminum	Al_2O_3 – aluminum oxide	9.3
Ta – tantalum	Ta_2O_5 – tantalum pentoxide	26
Nb – niobium	Nb_2O_5 – niobium pentoxide	42

Tab. 3.2: Relative permittivity of different dielectrics of electrolytic capacitors

Specification of the relative permeability of the different dielectrics illustrates that the capacitance per volume to be attained strongly depends on the electrode material used. This is because a higher permittivity allows a higher capacitance to be achieved with the same capacitor volume. This means that higher capacitance values can be achieved with tantalum or niobium electrolytic capacitors of the same package size than with aluminum electrolytic capacitors, or the same capacitance values with a smaller package size.

The general correlations of an electrolytic capacitor are now illustrated with the example of an aluminum electrolytic capacitor. The schematic structure of an aluminum electrolytic capacitor is as follows:

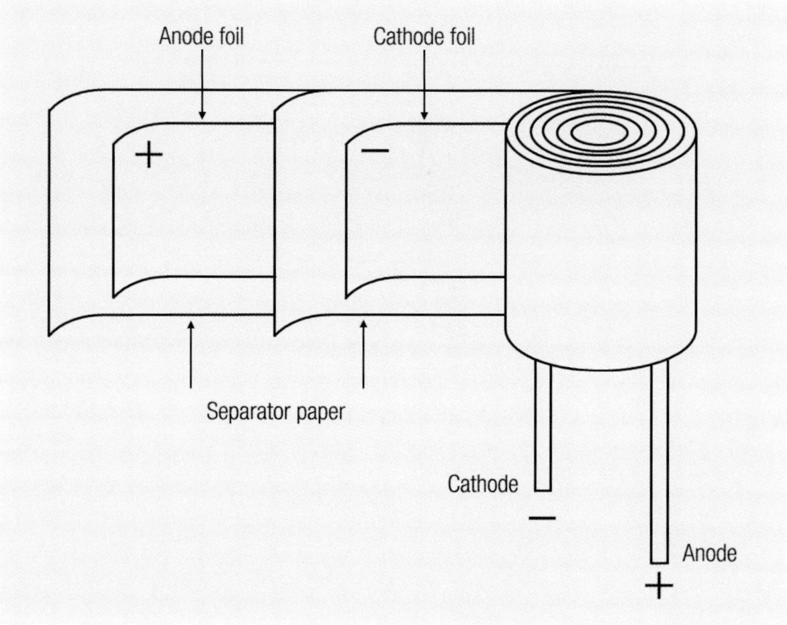

Fig. 3.6: Schematic structure of an aluminum capacitor

Aluminum electrolytic capacitor

The aluminum electrolytic capacitor is also a wound capacitor. The metal films are wound with an intermediate paper film in a similar way to those with a paper film capacitor. The connection pins are crimped onto the respective metal film. The paper layer (also known as separator) shown here is finally completely soaked with electrolyte and besides retaining the electrolyte also has the function of ensuring the separation between the metal films.

The following figure presents the layer structure within an aluminum-electrolyte capacitor in detail.

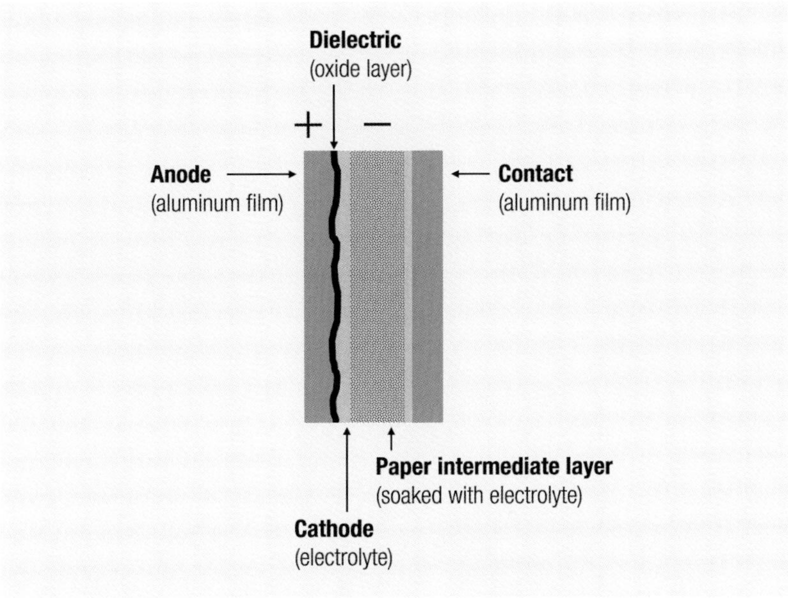

Fig. 3.7: Principle construction of an aluminum electrolytic capacitor with wet electrolyte

Often the cathode film is mistaken for the actual electrode, but this only establishes the electrical connection with the electrolyte. The electrolyte in electrolytic capacitors is the actual cathode and not the cathode film.

The oxide layer is generated with an electrochemical process, known as forming. The thickness of the oxide layer produced is proportional to the voltage applied.

The anode film is made rougher or smoother depending on the intended use. Here the surface is specially roughened with etching processes. Its surface structure differs depending on whether a high capacitance or high voltage stability is required.

The oxide layer has semiconductor properties, is only insulating in one direction and therefore only acts as a dielectric in one direction. With electrolytic capacitors, attention always has to be paid that the right polarity is observed. In general, electrolytic capacitors have one polarity and are damaged or destroyed if the polarity is reversed. The reason is that incorrect polarity causes reforming. Gases are formed and there is an increase in temperature in the capacitor. This can cause the electrolytic capacitor to swell or even burst. This is also the reason why wet electrolytic capacitors, for instance, have a vent fitted in the form of perforation on the component head. This can prevent the electrolytic capacitor from bursting in the event of reversing the polarity and the electrolyte going over the board. Given positive pressure in the electrolytic capacitor, the valve opens and the gas can escape.

The following are examples of a THT and a snap-in electrolytic capacitor. It is shown how the polarity is indicated on the capacitor for the user:

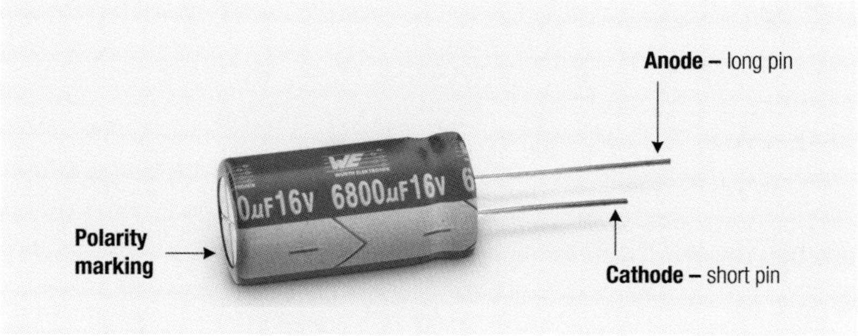

Fig. 3.8: Polarity of a THT electrolytic capacitor

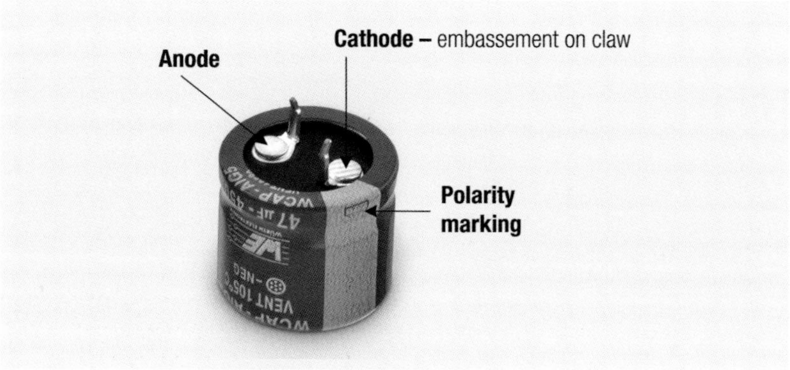

Fig. 3.9: Polarity of a THT Snap-in electrolytic capacitor

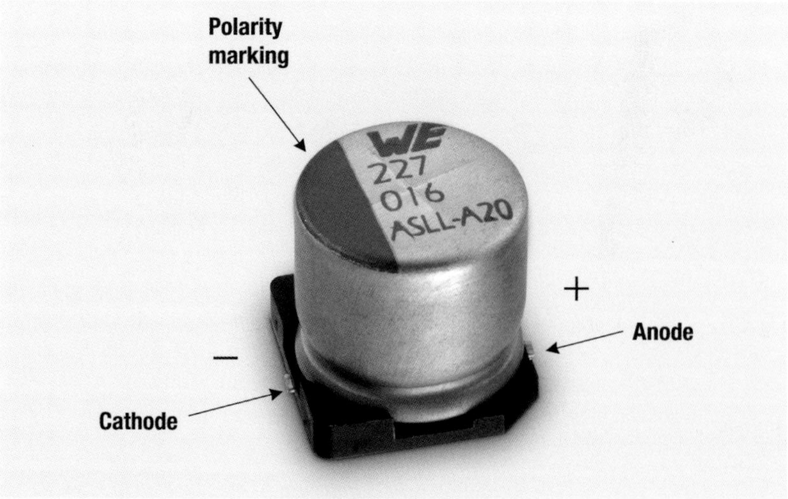

Fig. 3.10: Polarity of an SMD electrolytic capacitor

There are also non-polar or bipolar electrolytic capacitors, however. These have special constructions and are less widespread than polar versions. With bipolar electrolyte capacitors, there is an oxide layer on both electrode films and the construction principle is equivalent to series connection of two capacitors.

The internal series connection of a bipolar electrolytic capacitor reduces the possible volume capacitance to around half that of a similarly sized polar electrolytic capacitor with the same rated voltage.

The circuit symbols for polar and non-polar electrolytic capacitors differ from the general circuit symbol for capacitors:

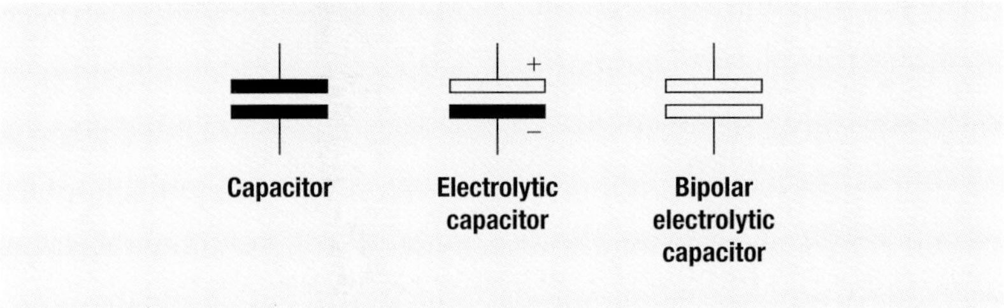

Fig. 3.11: Circuit symbol for electrolytic capacitors

There are a large number of different packages of electrolytic capacitors. Electrolytic capacitors are available with connections for SMT, THT, Snap-in and screw versions.

Fig. 3.12: Overview of the different electrolytic capacitor packages

3.4 Ceramic capacitors

Ceramic capacitors

As the name suggests, ceramic materials form the dielectric in ceramic capacitors. The ceramic capacitor can be structured as a homogeneous block (single layer capacitor) or in many layers (multilayer capacitor) that give rise to a monolithic ceramic body.

The single layer capacitor is found in high voltage applications and also in AC voltage applications. Multilayer capacitors are mainly configured as SMD components. The usual chip design is known as a multilayer ceramic capacitor and this led to the abbreviation MLCC common in industry.

MLCCs are the designs that predominate in the world market with quadrillions of units. The capacitor electrodes for this design are internal metalization layers. These are contacted to a terminal attached to the end face on the ceramic body.

The structure of the ceramic capacitors enables very low inductance designs. The ceramics used as the dielectric in ceramic capacitors are divided into various classes. The existing variants are in classes 1 to 4. The individual classes essentially differ in their relative permittivity ε_r, which influences the possible capacitance, the dependency towards temperature, their aging behavior and the dielectric losses. The differences in dependency of individual ceramics are illustrated with examples in Chapter 2.1 Capacitance of a capacitor.

It should always be borne in mind that although ceramic capacitors are suitable for achieving small packages and with SMD versions the mounting process can be considerably simplified, ceramics are more sensitive to mechanical influences than other designs. This has to be considered in manufacturing and in the application itself.

The individual ceramic classes are detailed in the following chapters.

3.4.1 Class 1 (low dielectric constant) ceramic capacitors

Class 1 ceramics

Class 1 (low dielectric constant) have a lower relative permittivity ($\varepsilon_r < 500$) than class 2 and higher ceramics. Capacitances in the range 0.5 pF to 0.47 µF can therefore be achieved with class 1 ceramics. However, ceramic capacitors of this class display linear temperature behavior (TC is linear or ideally TC=0). The linear temperature behavior makes them markedly different from ceramics of other classes.

Furthermore, class 1 ceramic capacitors have a high insulation resistance R_{INS}, lower self-discharge and there are therefore lower losses.

Precise capacitance values can be produced with class 1 ceramics, such that capacitance tolerances of ±1% to ±10% are achievable. Class 1 ceramics are always preferable to other ceramics in applications in which a very precise capacitance value is necessary (also in continuous operation) or if the application is subject to strong temperature fluctuations.

Ceramics are coded according to EIA-RS-198 or IEC60384-21. Codes consisting of three characters are used for identifying the ceramic capacitor in both systems. The breakdown of the codes is shown in the following tables:

EIA-RS-198 coding for class 1 ceramic capacitors					
1st character		2nd character		3rd character	
Letter	Temperature coefficient TC [ppm/°C]	Number	Multiplicator	Letter	Tolerance of temperature coefficient TC [ppm/°C] (for 28–85°C)
C	0.0	**0**	–1	**G**	±30
B	0.3	**1**	–10	**H**	±60
L	0.8	**2**	–100	**J**	±120
A	0.9	**3**	–1000	**K**	±250
M	1.0	**4**	+1	**L**	±500
P	1.5	**6**	+10	**M**	±1000
R	2.2	**7**	+100	**N**	±2500
S	3.3	**8**	+1000		
T	4.7				
V	5.6				
U	7.5				

Tab. 3.3: Excerpt from EIA-RS-198 coding for class 1 ceramic capacitors

IEC 60384-21 coding for class 1 ceramics			
Coding			
Identifier	Temperature coefficient TC [ppm/°C]	Tolerance of the temperature coefficient TC [ppm/°C]	Equivalent EIA-RS-198 coding
P100	100	±30	M7G
NP0	0	±30	C0G
N33	–33	±30	S2G
N75	–75	±30	U1G
N150	–150	±60	P2H
N220	–220	±60	R2H
N330	–330	±60	S2H
N470	–470	±60	T2H
N750	–750	±120	U2J
N1000	–1000	±250	M3K
N1500	–1500	±250	P3K

Tab. 3.4: Excerpt from IEC 60384-21 coding for class 1 ceramic capacitors

In the IEC identifier, N stands for negative and P for positive.

Probably the best known member of the class 1 ceramics is the NP0 or C0G, as it has a temperature coefficient TC=0 and therefore always delivers stable capacitance over the permissible temperature range. In reality, however, there is also a ±30 ppm/°C deviation.

3.4.2 Class 2 (high dielectric constant) ceramic capacitors

Class 2 ceramics

Class 2 ceramics (high dielectric constant) have higher achievable capacitance values than class 1 ceramics. But they are also associated with higher tolerances and higher losses. What is most significant is that these capacitors have non-linear temperature dependency. For this reason, no temperature coefficient TC is specified for these capacitors, instead the capacitance dependency behavior is specified to cover the entire permissible temperature range.

A further factor to be noted is that class 2 ceramics are subject to aging behavior, which has to be taken into consideration in the circuit design. The capacitor's available capacitance drops with increasing age. The capacitance available also drops with rising voltage (see chapter 2.1 Capacitance of a capacitor).

Class 2 is not suitable for applications which required a stable or precise capacitance value.

Like class 1, the identification of class 2 ceramics is determined by the EIA and IEC standards.

Identification in accordance with EIA-RS-198 is presented as follows.

EIA-RS-198 coding for class 2 ceramic capacitors					
1st character		**2nd character**		**3rd character**	
Letter	**Lower temperature limit**	**Number**	**Upper temperature limit**	**Letter**	**Capacitance change over the permissible temperature range**
X	−55 °C	**2**	+45 °C	**A**	±1.0%
Y	−30 °C	**4**	+65 °C	**B**	±1.5%
Z	+10 °C	**5**	+85 °C	**C**	±2.2%
		6	+105 °C	**D**	+3.3%
		7	+125 °C	**E**	+4.7%
		8	+150 °C	**F**	+7.5%
		9	+200 °C	**P**	±10%
				R	±15%
				S	±22%
				T	+22/−33%
				U	+22/−56%
				V	+22/−82%

Tab. 3.5: Excerpt from EIA-RS-198 coding for class 2 ceramic capacitors

For example, a X7R ceramic has a temperature range −55 °C to +125 °C and a permissible capacitance tolerance of ±15%.

As illustrated in the table, the EIA identifier is a three-character alphanumeric code. The code consists of a letter for the lower temperature limit, a number for the upper temperature limit and a letter for the permissible capacitance change.

Typical members of the class 2 ceramic capacitors are:

- X8R – $\Delta C/C_0$ = ±15%; perm. temperature range: −55 °C … +150 °C
- X7R – $\Delta C/C_0$ = ±15%; perm. temperature range: −55 °C … +125 °C
- X7S – $\Delta C/C_0$ = ±22%; perm. temperature range: −55 °C … +125 °C
- X5R – $\Delta C/C_0$ = ±15%; perm. temperature range: −55 °C … +85 °C
- Y5V – $\Delta C/C_0$ = +22/−82%; perm. temperature range: −30 °C … +85 °C
- Z5U – $\Delta C/C_0$ = +22/−56%; perm. temperature range:+10 °C … +85 °C

There is also coding in accordance with IEC 60384-22:

IEC 60384-22 coding for class 2 ceramics					
Character 1 and 2			**Character 3**		
Number & letter	**Maximum capacitance change $\Delta C/C_0$ at U = 0**	**Maximum capacitance change $\Delta C/C_0$ at U = UN**	**Number**	**Lower temperature limit**	**Upper temperature limit**
2B	±10%	+10%/–15%	**1**	–55 °C	+125 °C
2C	±20%	+20%/–30%	**2**	–55 °C	+125 °C
2D	+20%/–30%	+20%/–40%	**3**	–40 °C	+125 °C
2E	+22%/–56%	+22%/–70%	**4**	–25 °C	+125 °C
2F	+30%/–80%	+30%/–90%	**5**	–10 °C	+70 °C
2R	±15%	+15%/–40%	**6**	+10 °C	+85 °C
2X	±15%	+15%/–25%			

Tab. 3.6: Excerpt from IEC 60383-22 coding for class 2 ceramic capacitors

So an X7R ceramic is equivalent to a 2X1 ceramic. The EIA notation is common today for class 1 and 2 ceramics.

3.4.3 Class 3 and class 4 ceramic capacitors

Class 3 and class 4 capacitors are so-called junction capacitors. These are hardly of any importance today, but were more widespread in the 1990s.

These ceramics have a high relative permittivity (ε_r up to approx. 50000), although they show even more pronounced dependency behavior in terms of temperature, voltage and aging than class 2 capacitors. Moreover, it was only possible to achieve voltage ranges of 60–100 V with these ceramics.

Abc of Capacitors

Basic principles, characteristics and capacitor types

Indices

4 Indices

4.1 Keyword Index

A

C

D

E

F

T

4.2 Bibliography

No.	Title	Author/Publisher	Publishing year	Edition	ISBN-10	ISBN-13
1	Elemente der angewandten Elektronik: Kompendium für Ausbildung und Beruf	Erwin Böhmer, Dietmar Erhardt, Wolfgang Oberschelp"	2010	16.	3834805432	978-3834805430
2	Stand der Technik und Anwendung von Superkonden-satoren	Mustapha Jamal	2009	1.	3640523962	978-3640523962
3	Kondensatorenkunde für Elektroniker	Leucht, Kurt	1981		3772314910	–
4	Filmkondensatoren. Bauarten, Technologien und Anwend-ungen	Loth, Hans	1990	1.	3478930464	978-3478930468
5	Leitfaden für die Anwendung von Aluminium-Elektrolyt-Kondensatoren	Beuth Verlag	–	–	3410144714	978-3410144717
6	Embedded Electronics 1: Passive Bauelemente	Matthes Wolfgang	2008	1.	3895761842	978-3895761843
7	Elektronik-Fibel	Schnabel, Patrick	2010	5.	3831145903	978-3831145904
8	Taschenbuch der Elektrotechnik und Elektronik	Lindner, Helmut (Autor), Brauer, Harry (Autor), Lehmann, Constans (Autor)	2008	9.	3446414584	978-3446414587
9	Bauelemente – Elektronik 2	Klaus Beuth	2003	17.	380231957-5	–
10	Grundlagen der Elektrotechnik	Gert Hagmann	2001	9.	3891046618	–
11	Grundlagen und Bauelemente der Elektrotechnik	Heinz-Josef Bauckholt	2013	1.	3446432469	978-3446432468
12	Elektronische Bauelemente kurz erklärt. Technologien. Physikalische Grundlagen. Anwendungen	Dieter Frölich	1991	1.	3800915804	978-3800915804
13	Basiswissen Elektronik, Bd.2, Bauelemente	Dietmar Benda	1999	3.	3800718367	978-3800718368
14	Elektronische Bauelemente: Funktion, Grundschaltungen, Modellierung mit SPICE	Michael Reisch	2006	1.	3540340149	978-3540340140
15	Das große Werkbuch Elektronik	Dieter Nührmann	1994	6.	3772365469	978-3772365461
16	Das komplette Werkbuch Elektronik	Dieter Nührmann	2002	2.	3772365264	978-3772365263
17	Elektronik für Ingenieure und Naturwissenschaftler	Ekbert Hering, Klaus Bressler, Jürgen Gutekunst	2005	5.	3540243097	978-3540243090

No.	Title	Author/Publisher	Publishing year	Edition	ISBN-10	ISBN-13
18	Widerstände, Kondensatoren, Spulen und ihre Werkstoffe	O. Zinke, H. Seither	1982	2.	3540113347	978-3540113348
19	Der leichte Einstieg in die Elektrotechnik	Bo Hanus	2004	1.	3772359051	978-3772359057
20	Capacitor Handbook	Cletus J. Kaiser	1998	2.	–	978-0-9628525-3-4
21	Capacitors: Theory, Types and Applications	Alexander L. Schulz	2010	1.	1616689722	9781616689728
22	Electrical power Capacitor: Design and Manufacture	D. M. Tagare	2002	3.	0-07-463594-8	978-0-07-463594-0
23	Capacitors: Technology and Trends	R. P. Deshpande	2013	1.	1259063887	978-1259063886
24	Capacitors: Theory of Operation, Behavior and Safety Regulations	Kristofeer N. Müller	2013	1.	1624175856	978-1624175855
25	Encyclopedia of Electronic Components Volume 1: Resistors, Capacitors, Inductors, Switches, Encoders, Relays, Transistors	Charles Platt	2012	1.	1449333893	978-1449333898
26	Thin-Film Capacitors for Packaged Electronics	Jain Pushkar, Eugene J. Rymaszewski	2003	1.	140207705X	978-1402077050
27	Ceramic Dielectrics and Capacitors (Electrocomponent Science Monographs)	J.M.Herbert	1985	6.	2881240453	978-2881240454
28	Electrical Engineering 101, Third Edition: Everything You Should Have Learned in School ... but Probably Didn't	Darren Ashby	2011	3.	0123860016	978-0123860019
29	Elektro T: Grundlagen der Elektrotechnik	Wolfgang Bieneck	2003	4.	3-7782-4900-2	–

No.	Title	Author/Publisher	Link
1	Werkstoffe der Elektrotechnik	A.Thiede	http://groups.uni-paderborn.de/hfe/lehre/Script_wks_4.pdf
2	Frequenzgang der komplexen Permittivtät	J. Prochotta	http://et.fh-duesseldorf.de/e_einr/d_labore/physik/Permittivitaet.PDF